2nd International Conference on Physics for Sustainable Development and Technology (ICPSDT-2017)

Journal of Physics: Conference Series
Volume 1086

Chittagong, Bangladesh
10 – 11 December 2017

ISBN: 978-1-5108-7513-5
ISSN: 1742-6588

Printed from e-media with permission by:

Curran Associates, Inc.
57 Morehouse Lane
Red Hook, NY 12571

CURRAN ASSOCIATES INC. proceedings.com

Some format issues inherent in the e-media version may also appear in this print version.

Printed by Curran Associates, Inc. (2018)

Institute of Physics
Dirac House, Temple Back
Bristol BS1 6BE UK

Phone: 44 1 17 929 7481
Fax: 44 1 17 920 0979

techtracking@iop.org

Additional copies of this publication are available from:

Curran Associates, Inc.
57 Morehouse Lane
Red Hook, NY 12571 USA
Phone: 845-758-0400
Fax: 845-758-2633
Email: curran@proceedings.com
Web: www.proceedings.com

Table of contents

Volume 1086

2nd International Conference on Physics for Sustainable Development and Technology (ICPSDT-2017)

10-11 December 2017, Bangladesh

Accepted papers received: 9 August 2018
Published online: 13 September 2018

Preface

Papers

IOP Conf. Series: Journal of Physics: Conf. Series **1086** (2018) 011001 doi:10.1088/1742-6596/1086/1/011001

PREFACE

International Conference on Physics for Sustainable Development & Technology (ICPSDT) is a bi-annual conference which offers a place to present and discuss innovative and ongoing experimental and theoretical research works and/or their applications.

2nd ICPSDT-2017 held on 10-11 December, 2017 at Chittagong, Bangladesh is following the success and huge response of 1st ICPSDT-2015 held on 19-20 August, 2015 at Chittagong University of Engineering & Technology (CUET), Chittagong, Bangladesh and a day-long national conference on "Materials Science & Technology for Sustainable Development: Bangladesh Perspective" held on 8 November, 2009 organized by the Department of Physics, CUET, is certainly been increased keenness for the 2nd ICPSDT-2017.

The conference program encompasses nine technical sessions, including about 91 contributory talks and 59 posters covering most of the fundamental and applied physics and engineering topics as well. After rigorous peer review, best 15 papers were accepted for IOP, JPCS and 9 papers were for International Journal of Integrated Science & Technology (IJIST) published by Faculty of Engineering & Technology, CUET, Chittagong, Bangladesh. This proceeding covers a wide range of topic in Physics & its applications.

We would like to thank to all authors and coauthors for their contributions. Many thanks are endorsed to the reviewers for their competent evaluation of the submitted articles. Thanks extended to the members of organizing committee, technical committee, advisory committee and other committee members for their outstanding involvement to organize ICPSDT-2017 a fruitful event. Appreciation also goes to the sponsored partners: Bangladesh Advance Technologies (BAT), Sensor International, OMC, Elite paint, and Multi-tech Engineering for their enormous support. Last but not least, CUET authority and University Grants Commission (UGC) of Bangladesh have given tremendous logistic and financial support to make this event a successful one.

We sincerely hope that ICPSDT will be an excellent forum for discussions that will lead to fruitful research, studies and other new scientific ideas and engineering products that will be beneficial to the society and the industry.

Organizing Committee

- **Prof. Dr. Faruque-Uz-Zaman Chowdhury**, Dept. of Physics, CUET **(Chair)**
- **Prof. Dr. Md. Abdur Rashid**, Dept. of Physics, CUET
- **Prof. Dr. Md. Mohi Uddin**, Dept. of Physics, CUET
- **Mrs. Nusrat Jahan**, Dept. of Physics, CUET
- **Mrs. Mosammat Arjumanara Bagum**, Dept. of Physics, CUET
- **Mr. Md. Ashraf Ali**, Dept. of Physics, CUET
- **Miss Syeda Karimunnesa**, Dept. of Physics, CUET
- **Dr. Md. Mukter Hossain**, Dept. of Physics, CUET
- **Mr. Shaiful Kabir**, Dept. of Physics, CUET
- **Dr. Animesh Kumer Chakraborty**, Dept. of Physics, CUET **(Treasurer)**
- **Mr. H. M. A. R. Maruf**, Dept. of Physics, CUET **(Joint Member Secretary)**
- **Prof. Dr. Mohammad Belal Hossen**, Dept. of Physics, CUET **(Member Secretary)**

Technical Committee

- **Prof. Dr. Md. Abdur Rashid**, Dept. of Physics, CUET **(Chair)**
- **Prof. Dr. Faruque-Uz-Zaman Chowdhury**, Dept. of Physics, CUET
- **Prof. Dr. Mohammad Belal Hossen**, Dept. of Physics, CUET
- **Mrs. Nusrat Jahan**, Dept. of Physics, CUET
- **Mr. H. M. A. R. Maruf**, Dept. of Physics, CUET
- **Mr. Md. Ashraf Ali**, Dept. of Physics, CUET
- **Prof. Dr. A.K.M. Akther Hossain**, , Dept. of Physics, BUET
- **Prof. Dr. Md. Moazzem Hossain Miah**, , Dept. of Physics, Chittagong University
- **Prof. Dr. Md. Rafiqul Islam**, Dept. of Physics, Chittagong University
- **Prof. Dr. Rezaul Huque Khan**, Independent University, Chittagong
- **Prof. Dr. Saleh Hasan Naqib**, Dept. of Physics, Rajshahi University
- **Prof. Dr. Mayeen Uddin Khandaker**, UM, Malaysia
- **Dr. Md. Abdul Gafur**, BCSIR, Dhaka
- **Dr. A.K.M. Zakaria**, Planning & Development Division, BAEC HQ, Dhaka
- **Dr. Md. Shuza Uddin**, CSO, AERE, Dhaka
- **Dr. Md. Mukter Hossain,** Dept. of Physics, CUET **(Joint Member Secretary)**
- **Prof. Dr. Md. Mohi Uddin,** Dept. of Physics, CUET **(Member Secretary)**

IOP Conf. Series: Journal of Physics: Conf. Series **1086** (2018) 011002 doi:10.1088/1742-6596/1086/1/011002

Peer review statement

All papers published in this volume of *Journal of Physics: Conference Series* have been peer reviewed through processes administered by the proceedings Editors. Reviews were conducted by expert referees to the professional and scientific standards expected of a proceedings journal published by IOP Publishing.

IOP Conf. Series: Journal of Physics: Conf. Series **1086** (2018) 012001 doi:10.1088/1742-6596/1086/1/012001

Positron acoustic solitary waves in an inhomogeneous multi-component plasma

M S Alam[1,3], M G Hafez[1, 3], N C Roy[2], M R Talukder[2] and M H Ali[1]

[1]Department of Applied Mathematics, University of Rajshahi, Rajshahi-6205, Bangladesh
[2]Plasma Science and Technology Lab, Department of Applied Physics and Electronic Engineering, University of Rajshahi, Rajshahi-6205, Bangladesh.
[3]Department of Mathematics, Chittagong University of Engineering and Technology, Chittagong-4349, Bangladesh.

*E-mail: alam21nov_2016@yahoo.com, **E-mail: mrtalukder@ru.ac.bd

Abstract. The theoretical investigations have been made for the propagation of positron acoustic solitary waves (PASWs) in a weakly inhomogeneous plasma composing immobile positive ions, mobile cold positrons, and superthermal hot positrons and electrons. The Korteweg-de Varies (KdV) and modified KdV (mKdV) equations with variable coefficients are derived using the appropriate coordinate transformation and the reductive perturbation method (RPM). The effects of positron concentration, temperature ratios for hot positrons and electrons, hot to cold positrons density ratio, electron to cold positron density ratio, ion to cold positron density and population of hot electrons as well as positrons superthermality on the nonlinear propagation of PASWs are examined to understand the local electrostatic disturbances. It is also found that the presence of superthermal (kappa distributed) hot positrons and hot electrons significantly modify the basic features of PASWs. The critical values for hot positrons and cold positrons also play a vital role in the formation of only compressive PASWs in the plasmas.

1. Introduction

It is well known that the electron–positron–ion (e-p-i) plasmas are existed in astrophysical, space as well as laboratory plasmas, especially in ionosphere [1], auroral acceleration regions [2], solar wind [3], quasar and pulsar magnetosphere [4], active galactic nuclei [4], polar cup of fast rotating neutron stars [5], semiconductor plasmas [6], intense laser fields [7] and so on. However, it is not predominantly easy to the researchers for the production of astrophysical or space like plasmas in the laboratory. One may be studied by finding the basic characteristics of plasmas for understanding the physical issues involved in the plasmas. Further, one may frequently encounter nonlinear collective influences to the plasmas, which cannot appropriately be studied without tedious mathematical techniques. The dispersion or dissipation along with nonlinearities provide several types of consistent structures, such as solitary waves, shock structures, double layer (DL), vortices, etc. which play important roles for understanding the physical phenomena from both the theoretical and experimental point of views. On the other hand, space plasmas with an excess of superthermal electrons or positrons are generally characterized by a long tail in the high energy region, which can be studied by considering generalized Lorentzian or kappa distributions [8, 9].The presence of a significant number of superthermal particles follow kappa distribution, but not the Maxwellian one, can significantly change the rate of resonant energy transfer between particles and plasma waves [10]. Besides, C. M. Surko et al. [11] and Abdullah et al. [12] have shown that the cold positrons are now available as new sources in laboratory plasmas, such as semiconductor plasmas and intense laser fields [7]. Therefore, the studies of nonlinear wave propagation in the e-p-i superthermal plasmas become one of essential aspects in recent years due to its broad applications and potentiality as mentioned earlier. On the other hand, many authors [13-20] have studied the nonlinear positron acoustic waves (PAWs) in different e-p-i homogenous plasmas considering various types of plasma assumptions. Recently, Alam et al. [20] have studied thepositron acoustic (PA) solitary waves (SWs)and double layer (DL) by deriving Korteweg-de Varies (KdV), modified KdV (mKdV) and Gardner equations in an unmagnetized homogeneous plasma system, which is composed of immobile positive ions, mobile cold positrons and superthermal hot electrons as well as positrons. They have found that the kappa distributed hot electrons and positrons significantly modify amplitudes, widths, polarity, and phase speed of PASWs and DLs in the plasmas.

In most of the studies [13-29] have investigated the basic properties of electrostatic waves in the homogeneous plasmas. But, the influence of plasma inhomogeneity on the propagation of electrostatic acoustic waves is another significant problem from various points of view. For instance, one may consistently encounter inhomogeneous plasmas under actual conditions both in the laboratory as well as space plasmas. Inhomogeneity may stem from density gradient or temperature or it could be due to the magnetic field in space and they are more pronounced closer to the edges and the boundaries of the system. The characteristics such as propagation, collision, and reflection of solitons etc., are significantly influenced under such conditions. Only a few authors [30–34] have been investigated that the propagation of ion acoustic solitary waves in an inhomogeneous plasma considering different plasma assumptions. Mowafy and Moslem [33] have studied the propagation of the IAWs in an inhomogeneous plasma comprised warm negative and positive ions with different masses and isothermal electrons**.** It is best to our knowledge no one investigate the nonlinear propagation of PASWs in an inhomogeneous plasma consisting of immobile positive ions, mobile cold positron, and superthermal hot positrons as well as electrons. Due to the significance of the problems related to the astrophysical, space as well as laboratory plasmas, the effects of plasma parameters on the nonlinear propagation of PASWs in a weakly unmagnetized inhomogeneous plasma consisting of immobile positive ions, mobile cold positrons, superthermal hot positrons as well as electrons are investigated. Thus, the paper is organized as follows: the theoretical model and derivations of KdV and mKdV equations in inhomogeneous plasmas with analytical solutions are presented in Section 2. The results and discussion are described in Section 3. Finally, the conclusion is drawn in section 4.

2. Model equations

A fully ionized, collisionless, unmagnetized four-component plasma system consisting of immobile positive ions, mobile cold positron, and kappa distributed hot positrons, as well as electrons, is considered. At equilibrium, the charge neutrality condition yields $n_{e0} = n_{pc0} + n_{ph0} + n_{i0}$, where $n_{i0}, n_{e0}$and $n_{pc0}(n_{ph0})$ are the unperturbed ion number density, electron number density and cold (hot) positron number density, respectively. The concentrations of hot positrons (n_{ph}) and hot electrons (n_e) taking the kappa distributed function into account are obtained [20] as

$$n_{ph} = n_{ph0}\left[1+\frac{e\phi}{k_BT_{ph}\left(\kappa_p-\frac{3}{2}\right)}\right]^{-\kappa_p+\frac{1}{2}} \text{and} n_e = n_{e0}\left[1-\frac{e\phi}{k_BT_e\left(\kappa_e-\frac{3}{2}\right)}\right]^{-\kappa_e+\frac{1}{2}},$$

where T_{ph} and T_e are respectively the temperatures of hot positrons and hot electrons, and κ_p and $\kappa_e$are respectively the superthermal parameters of hot positrons and hot electrons and k_B is the Boltzmann constant.In order to study the nonlinear dynamics of PASWs in the plasma system [20] taking in homogeneity into account, the normalized hydrodynamic fluid equations can be written as

$$\frac{\partial n_{pc}}{\partial t}+\frac{\partial}{\partial x}(n_{pc}u_{pc})=0\,, \tag{1}$$

$$\frac{\partial u_{pc}}{\partial t}+u_{pc}\frac{\partial u_{pc}}{\partial x}=-\frac{\partial \phi}{\partial x}\,, \tag{2}$$

$$\frac{\partial^2\phi}{\partial x^2}=-n_{pc}-\mu_{ph}\left(1+\frac{\sigma_1\phi}{\kappa_p-\frac{3}{2}}\right)^{-\kappa_p+\frac{1}{2}}+\mu_e\left(1-\frac{\sigma_2\phi}{\kappa_e-\frac{3}{2}}\right)^{-\kappa_e+\frac{1}{2}}-\mu_i, \tag{3}$$

where, n_{pc} is the cold positron number density normalized by its equilibrium value n_{pc0}, u_{pc} is the cold positron fluid speed normalized by positron acoustic speed $C_{pc}=\left(k_BT_{ef}/m_p\right)^{1/2}$, ϕ is the electrostatic potential normalized by k_BT_{ef}/e, $\sigma_1=T_{ef}/T_{ph}$, $\sigma_2=T_{ef}/T_e$, $\mu_{ph}=n_{ph0}/n_{pc0}$, $\mu_e=n_{e0}/n_{pc0}$, $\mu_i=n_{i0}/n_{pc0}$, $T_{ef}=T_eT_{ph}/(\mu_eT_{ph}+\mu_{ph}T_e)$ is the effective temperature, m_p is the positron mass and e is the magnitude of electron charge. The time and space variables are normalized by the plasma period $\omega_{pc}^{-1}=\left(m_p/4\pi n_{pc0}e^2\right)^{1/2}$ and Debye length $\lambda_{Dp}=\left(k_BT_{ef}/4\pi n_{pc0}e^2\right)^{1/2}$, respectively.

3. Derivation of KdV and mKdV equations with variable coefficients

Let us consider the stretched variables, which is spatially applicable for studying the nonlinear propagation of electrostatic PASWs in an inhomogeneous plasma system [33,35] as

$$\xi=\varepsilon^{1/2}\left(\int\frac{dx}{V(x)}-t\right),X=\varepsilon^{3/2}x, \tag{4}$$

where X is the time related coordinates, V is the speed of wave propagation and ε is a small parameter measuring the weakness of dispersion.
Now, expanding the perturbed quantities according to the well-known RPM as

$$\mathcal{H}=\mathcal{H}^{(0)}+\sum_{i=1}^{\infty}\varepsilon^i\mathcal{H}^{(i)}\,, \tag{5}$$

where $\mathcal{H}=(n_{pc}\quad u_{pc}\quad \phi)'$, $\mathcal{H}^{(0)}=\left(n_{pc}^{(0)}\quad u_{pc}^{(0)}\quad \phi^{(0)}\right)'$ and $\mathcal{H}^{(i)}=\left(n_{pc}^{(i)}\quad u_{pc}^{(i)}\quad \phi^{(i)}\right)'$, taking only spatial gradients $\frac{\partial\mathcal{H}^{(0)}}{\partial\xi}=0$, and $\frac{\partial V}{\partial\xi}=0$. Substituting Eqs. (4) and (5) into Eqs. (1)-(3) and collecting the quantities based on the different powers of ε, one can obtain a set of equations in terms of ε. The first order of ε gives the following relations:

$$\left.\begin{aligned}n_{pc}^{(1)}=\frac{n_{pc}^{(0)}\phi^{(1)}}{\left(V-u_{pc}^{(0)}\right)^2},u_{pc}^{(1)}=\frac{\phi^{(1)}}{\left(V-u_{pc}^{(0)}\right)},\\ n_{pc}^{(0)}/\left(V-u_{pc}^{(0)}\right)^2-R_1+2R_2\phi^{(0)}-3R_3\{\phi^{(0)}\}^2=0\end{aligned}\right\}, \tag{6}$$

where,

$$R_1 = \left\{ \frac{\mu_{ph}\left(\kappa_p - \frac{1}{2}\right)\sigma_1}{\kappa_p - \frac{3}{2}} + \frac{\mu_e\left(\kappa_e - \frac{1}{2}\right)\sigma_2}{\kappa_e - \frac{3}{2}} \right\},$$

$$R_2 = \left\{ \frac{\mu_{ph}\left(\kappa_p - \frac{1}{2}\right)\left(\kappa_p + \frac{1}{2}\right)\sigma_1^2}{2\left(\kappa_p - \frac{3}{2}\right)^2} - \frac{\mu_e\left(\kappa_e - \frac{1}{2}\right)\left(\kappa_e + \frac{1}{2}\right)\sigma_2^2}{2\left(\kappa_e - \frac{3}{2}\right)^2} \right\},$$

$$R_3 = \left\{ \frac{\mu_{ph}\left(\kappa_p - \frac{1}{2}\right)\left(\kappa_p + \frac{1}{2}\right)\left(\kappa_p + \frac{3}{2}\right)\sigma_1^3}{6\left(\kappa_p - 3/2\right)^3} + \frac{\mu_e\left(\kappa_e - \frac{1}{2}\right)\left(\kappa_e + \frac{1}{2}\right)\left(\kappa_e + \frac{3}{2}\right)\sigma_2^3}{6(\kappa_e - 3/2)^3} \right\}.$$

To the next higher order of ε gives the following equations

$$\frac{2}{\left(V - u_{pc}^{(0)}\right)^3}\phi^{(1)}\frac{\partial\phi^{(1)}}{\partial\xi} + \left(\frac{Vn_{pc}^{(0)}}{\left(V - u_{pc}^{(0)}\right)} + \frac{Vu_{pc}^{(0)}n_{pc}^{(0)}}{\left(V - u_{pc}^{(0)}\right)^2}\right)\frac{\partial\phi^{(1)}}{\partial X}$$
$$+ \left(\frac{V}{\left(V - u_{pc}^{(0)}\right)}\frac{\partial n_{pc}^{(0)}}{\partial X} + \frac{Vn_{pc}^{(0)}}{\left(V - u_{pc}^{(0)}\right)^2}\frac{\partial u_{pc}^{(0)}}{\partial X}\right)\phi^{(1)} = \left(V - u_{pc}^{(0)}\right)\frac{\partial n_{pc}^{(2)}}{\partial\xi} - n_{pc}^{(0)}\frac{\partial u_{pc}^{(2)}}{\partial\xi},$$

$$\frac{Vu_{pc}^{(0)}}{\left(V - u_{pc}^{(0)}\right)}\frac{\partial\phi^{(1)}}{\partial X} + \frac{V}{\left(V - u_{pc}^{(0)}\right)}\phi^{(1)}\frac{\partial u_{pc}^{(0)}}{\partial X} + V\frac{\partial\phi^{(1)}}{\partial X} + \frac{\partial\phi^{(2)}}{\partial\xi} + \frac{1}{\left(V - u_{pc}^{(0)}\right)^2}\phi^{(1)}\frac{\partial\phi^{(1)}}{\partial\xi}$$
$$= \left(V - u_{pc}^{(0)}\right)\frac{\partial u_{pc}^{(2)}}{\partial\xi},$$

$$-n_{pc}^{(2)} + R_1\phi^{(2)} - R_2\{\phi^{(1)}\}^2 + R_3\phi^{(0)}\{\phi^{(1)}\}^2 - R_2\phi^{(0)}\phi^{(2)} + R_3\phi^{(2)}\{\phi^{(0)}\}^2$$
$$= \frac{1}{\left(V - u_{pc}^{(0)}\right)^2}\frac{\partial^2\phi^{(1)}}{\partial\xi^2}.$$

Eliminating $n_{pc}^{(2)}$,$u_{pc}^{(2)}$ and $\phi^{(2)}$ from the above equations with the help of Eq. (6), one can derive the following nonlinear evolution equation with variable coefficients as

$$\frac{\partial\phi^{(1)}}{\partial X} + \frac{A_3}{A_2}\phi^{(1)}\frac{\partial\phi^{(0)}}{\partial X} + \frac{A_1}{A_2}\phi^{(1)}\frac{\partial\phi^{(1)}}{\partial\xi} + \frac{1}{V^2A_2}\frac{\partial^3\phi^{(1)}}{\partial\xi^3} = 0, \tag{7}$$

where $A_1 = R_2 - 2R_3\phi^{(0)} + \frac{6n_{pc}^{(0)}}{\left(V-u_{pc}^{(0)}\right)^4}$, $A_2 = \frac{2V^2n_{pc}^{(0)}}{\left(V-u_{pc}^{(0)}\right)^3}$and $A_3 = \frac{Vn_{pc}^{(0)}(V-3u_{pc}^{(0)})}{\left(V-u_{pc}^{(0)}\right)^3\{u_{pc}^{(0)}\}^2}$.
Eq. (7) is known as the KdV like the equation with variable coefficients. Using the variable transform$\phi^{(1)} = \varphi(\xi, X)e^{(-A_3/A_2)\phi^{(0)}}$, Eq. (7) can be converted to the following KdV equation:

$$\frac{\partial\varphi}{\partial X} + L\varphi\frac{\partial\varphi}{\partial\xi} + M\frac{\partial^3\varphi}{\partial\xi^3} = 0, \tag{8}$$

where $M = (1/V^2 A_2)$ and $L = (A_1/A_2)e^{(-A_3/A_2)\phi^{(0)}}$. However, the nonlinear coefficient functionally occurs based on the space of the plasma due to the sake of simplicity for mathematical development. The variations are assumed in significantly as compared to the scale length or it is supposed that all parameters are locally constant. Therefore, introducing a transformation ($\chi = \xi - U_0 X$), the solitary wave solutions of Eq. (8) can be obtained as

$$\varphi = \varphi_0 \text{sech}^2\left\{\frac{\chi}{W_0}\right\}, \tag{9}$$

where $\varphi_0 = (3U_0/L)$ and $W_0 = \sqrt{(4M/U_0)}$ are respectively the amplitudes and widths of KdV solitary waves andχ is the transformed coordinate with regards to a frame moving with constant velocity U_0. It is found that the amplitudes of the KdV solitary waves approach to infinity when $L \to 0$for $n_{pc}^{(0)} = \left(V - u_{pc}^{(0)}\right)^4 (R_2 - 2R_3\phi^{(0)})/6$, where the validity of the perturbation technique breaks down. In such case, the structures of PA solitary waves around the critical densities are studied considering the higher order nonlinear evolution equation in weakly inhomogeneous multi-components plasma. To do so, one needs to convert the stretched variables of Eq. (4) as

$$\xi = \varepsilon\left(\int \frac{dx}{V} - t\right), X = \varepsilon^3 x. \tag{10}$$

Using Eqs. (5) and (10) into Eqs. (1)-(3), one can obtain the same values of $n_{pc}^{(1)}$, $u_{pc}^{(1)}$ and the dispersion relation as mentioned in Eq. (6). To the next higher order of ε gives

$$-\left(V - u_{pc}^{(0)}\right)\frac{\partial n_{pc}^{(2)}}{\partial \xi} + n_{pc}^{(0)}\frac{\partial u_{pc}^{(2)}}{\partial \xi} + \frac{\partial}{\partial \xi}\left(n_{pc}^{(1)} u_{pc}^{(1)}\right) = 0 \tag{11}$$

$$-\left(V - u_{pc}^{(0)}\right)\frac{\partial u_{pc}^{(2)}}{\partial \xi} + u_{pc}^{(1)}\frac{\partial u_{pc}^{(1)}}{\partial \xi} + \frac{\partial \phi^{(2)}}{\partial \xi} = 0 \tag{12}$$

$$-n_{pc}^{(2)} + R_1\phi^{(2)} - 2R_2\phi^{(0)}\phi^{(2)} + 3R_3\{\phi^{(0)}\}^2\phi^{(2)} + [-R_2 + R_3\phi^{(0)}]\{\phi^{(1)}\}^2 = 0. \tag{13}$$

Solving Eqs. (11)-(13) with the help of Eq. (6), one can find the following relations

$$n_{pc}^{(2)} = \frac{3n_{pc}^{(0)}\{\phi^{(1)}\}^2}{2\left(V - u_{pc}^{(0)}\right)^4} + \frac{n_{pc}^{(0)}\phi^{(2)}}{\left(V - u_{pc}^{(0)}\right)^2}, u_{pc}^{(2)} = \frac{\{\phi^{(1)}\}^2}{2\left(V - u_{pc}^{(0)}\right)^3} + \frac{\phi^{(2)}}{\left(V - u_{pc}^{(0)}\right)}, \tag{14}$$

$$\left\{\frac{n_{pc}^{(0)}}{\left(V - u_{pc}^{(0)}\right)^2} - R_1 + 2R_2\phi^{(0)} - 3R_3\{\phi^{(0)}\}^2\right\}\phi^{(2)} + \frac{1}{2}\left\{R_2 - 2R_3\phi^{(0)} + \frac{6n_{pc}^{(0)}}{\left(V - u_{pc}^{(0)}\right)^4}\right\}\{\phi^{(1)}\}^2 = 0. \tag{15}$$

Finally, the following equations can be obtained considering the next higher order of ε:

$$-\left(V-u_{pc}^{(0)}\right)\frac{\partial n_{pc}^{(3)}}{\partial\xi}+n_{pc}^{(0)}\frac{\partial u_{pc}^{(3)}}{\partial\xi}+n_{pc}^{(1)}\frac{\partial u_{pc}^{(2)}}{\partial\xi}+u_{pc}^{(2)}\frac{\partial n_{pc}^{(1)}}{\partial\xi}+n_{pc}^{(2)}\frac{\partial u_{pc}^{(1)}}{\partial\xi}+u_{pc}^{(1)}\frac{\partial n_{pc}^{(2)}}{\partial\xi}$$
$$+V\left[n_{pc}^{(0)}\frac{\partial u_{pc}^{(1)}}{\partial X}+u_{pc}^{(1)}\frac{\partial n_{pc}^{(0)}}{\partial X}+n_{pc}^{(1)}\frac{\partial u_{pc}^{(0)}}{\partial X}+u_{pc}^{(0)}\frac{\partial n_{pc}^{(1)}}{\partial X}\right]=0 \quad (16)$$

$$-\left(V-u_{pc}^{(0)}\right)\frac{\partial u_{pc}^{(3)}}{\partial\xi}+\frac{\partial}{\partial\xi}\left[u_{pc}^{(1)}u_{pc}^{(2)}\right]+V\left(u_{pc}^{(0)}\frac{\partial u_{pc}^{(1)}}{\partial X}+u_{pc}^{(1)}\frac{\partial u_{pc}^{(0)}}{\partial X}\right)+\frac{\partial\phi^{(3)}}{\partial\xi}+\frac{\partial\phi^{(1)}}{\partial X}=0 \quad (17)$$

$$\frac{1}{V^2}\frac{\partial^2\phi^{(1)}}{\partial\xi^2}-n_{pc}^{(3)}=R_1\phi^{(3)}-2R_2\phi^{(0)}\phi^{(3)}-2R_2\phi^{(1)}\phi^{(2)}+R_3\{\phi^{(1)}\}^3+6R_3\phi^{(0)}\phi^{(1)}\phi^{(2)} \quad (18)$$

Eliminating $n_{pc}^{(3)}, u_{pc}^{(3)}$ and $\phi^{(3)}$ from Eqs. (16)-(18) using Eqs. (6), (14) and Eq. (15), one can derive the following evolution equation with variable coefficients as

$$\frac{\partial\phi^{(1)}}{\partial X}+\frac{A_5}{A_2}\phi^{(1)}\frac{\partial\phi^{(0)}}{\partial X}+\frac{A_4}{A_2}\{\phi^{(1)}\}^2\frac{\partial\phi^{(1)}}{\partial\xi}+\frac{1}{V^2A_2}\frac{\partial^3\phi^{(1)}}{\partial\xi^3}=0 \quad (19)$$

where $A_4=\frac{3}{2}\left\{\frac{n_{pc}^{(0)}}{\left(V-u_{pc}^{(0)}\right)^2}+\frac{2(1+n_{pc}^{(0)})}{\left(V-u_{pc}^{(0)}\right)^6}+2R_3\right\}, A_5=\frac{Vn_{pc}^{(0)}(2u_{pc}^{(0)}-V)}{\left(V-u_{pc}^{(0)}\right)^3\{u_{pc}^{(0)}\}^2}.$

To study the electrostatic potential structures, one can reduce Eq. (19) to the following form considering the transformation as $\phi^{(1)}=\varphi(\xi,X)e^{(-A_5/A_2)\phi^{(0)}}$:

$$\frac{\partial\varphi}{\partial X}+L_1\varphi^2\frac{\partial\varphi}{\partial\xi}+M\frac{\partial^3\varphi}{\partial\xi^3}=0, \quad (20)$$

where $L_1=(A_4/A_2)e^{(-A_5/A_2)\phi^{(0)}}$. Eq. (20) is the well known mKdV equation. Therefore, introducing a transformation ($\chi=\xi-U_0X$), the solitary wave solutions of Eq. (20) can be obtained as

$$\varphi=\varphi_1\mathrm{sech}\left\{\frac{\chi}{W_4}\right\}, \quad (21)$$

where $\varphi_1=(6U_0/L_1)$ and $W_4=\sqrt{(M/U_0)}$ are respectively the amplitudes and widths of mKdV PA solitary waves.

4. Results and Discussion

The physical issues concerned in the weakly inhomogeneous plasma (composing immobile positive ions, mobile cold positrons, and superthermal hot positrons as well as electrons) have investigated by driving the KdV and mKdV equations with variable coefficients. The parametric effects on the electrostatic PASWs are discussed considering the typical ranges $\mu_e=0.1-0.6$, $\sigma_1=1-6$, $\sigma_2=0.1-0.9$, $\kappa_e=1.8-2.2$ and $\kappa_p=3-100$ of the plasma parameters,which are consistent with space and laboratory plasmas [1-3, 7, 8].

Figs. 1(a)-(d) show the electrostatic PA KdV solitary waves for different values ofn_{pc}^0, μ_e,κ_e and σ_1 considering the remaining parameter constants. It is seen from Figs. 1 that the amplitudes of PASWs are decreasing with the increase of n_{pc}^0, μ_e, κ_e and σ_1.The nonlinear term (L) of KdV equation increases due to the enhancement of densities and spectral index kappa of electrons parameters, as a result of the amplitudes of the PASWs decrease. The driving force, due to the inertia of the positron, also decreases with increasing positron concentration, and as a result, the solitary waves are generated as well as the amplitudes of the PASWs decrease. It is also found that the KdV equation supports both compressive for $L < 0$ and rarefactive for $L > 0$ solitons depending on the plasma parameters in the inhomogeneous plasmas. Besides, the KdV equation does not support the PASWs in the weakly inhomogeneous plasmas at the critical unperturbed densities $n_{pc}^{(0)} = \left(V - u_{pc}^{(0)}\right)^4 (R_2 - 2R_3\phi^{(0)})/6$. In such case, the mKdV equation with variable coefficients is derived considering further higher order nonlinearity terms to investigate the PASWs around the critical densities. Figs. 2(a)-(d) display the effect of n_{pc}^0, κ_e, κ_p and σ_1 along with χ on the electrostatic PA mKdV solitary waves considering the remaining parameter constants. It is seen from Figs. 2 that the amplitudes of mKdV PASWs are increasing with the increase of n_{pc}^0, κ_e and κ_p, and decreasing with the increase of σ_1. It is interesting to note that the mKdV equation supports only the compressive solitons depending on the inhomogeneous plasma conditions. The results obtained in this manuscript are in good agreements with the findings of [20] in homogeneous plasmas. Thus, the parametric investigations on the considered inhomogeneous plasmas may be useful for understanding the physical properties of nonlinear electrostatic PA solitary waves in space plasmas [1-5] and laboratory plasmas [6, 7].

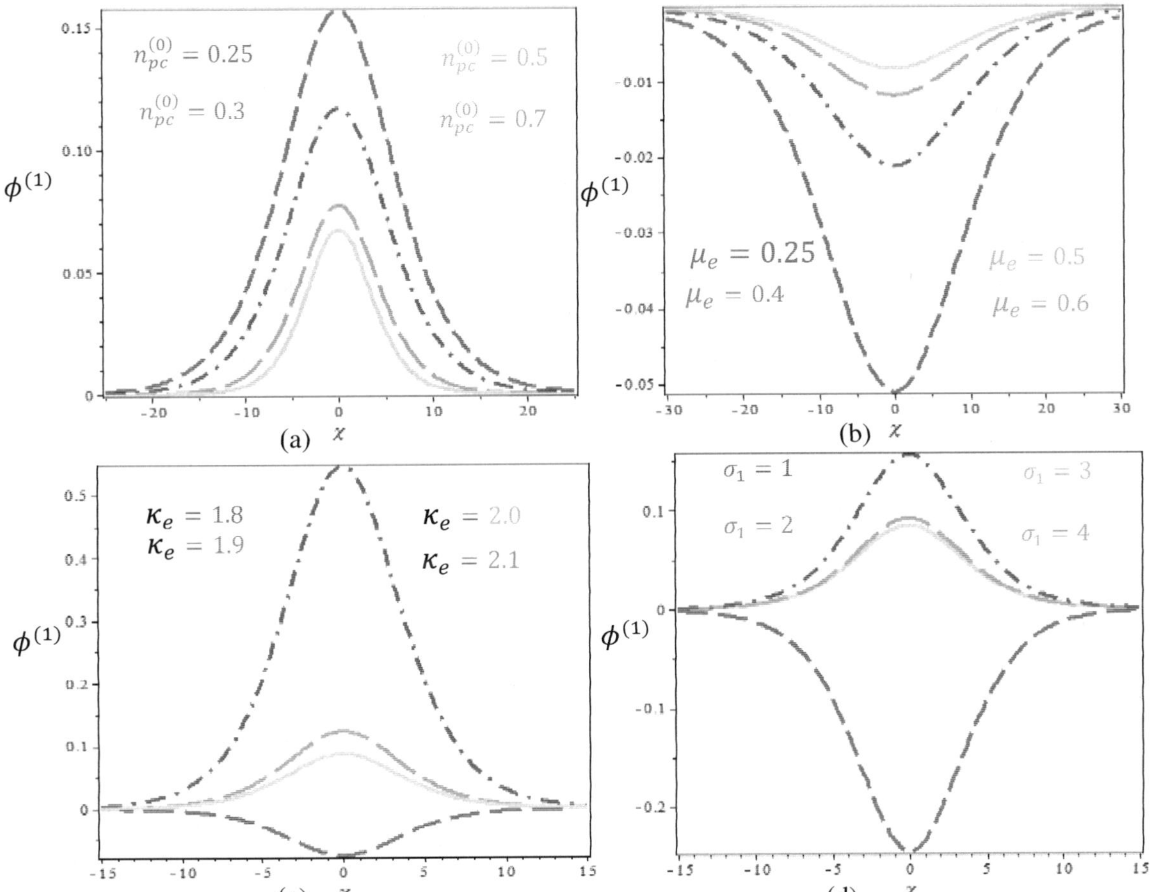

Figure 1. Effects of (a) $n_{pc}^{(0)}$ ($\kappa_e = 1.8$, $\mu_e = 0.2$, $\sigma_1 = 1$, $\mu_{ph} = 0.08$, $\kappa_p = 3$, $\sigma_2 = 0.5$,$u_{pc}^{(0)} = 0.6$, $V = 1.6$, $\phi^{(0)} = 0.1$, $U_0 = 0.02$) (b)μ_e with the same values of (a) but $\mu_{ph} = 0.14$ and $n_{pc}^{(0)} = 0.1$, (c)κ_e ($\mu_{ph} = 0.14$,$\mu_e = 0.6$, $\kappa_p = 3$, $\sigma_1 = 1$, $\sigma_2 = 0.5$, $u_{pc}^{(0)} = 0.6$, $n_{pc}^{(0)} = 0.7$, $V = 1.6$, $\phi^{(0)} = 0.5$, $U_0 = 0.02$) and (d) σ_1 with the same values of (c) but $\kappa_e = 2.1$ and $\kappa_p = 5$ on the electrostatic PA KdV solitary waves in weakly inhomogeneous plasma.

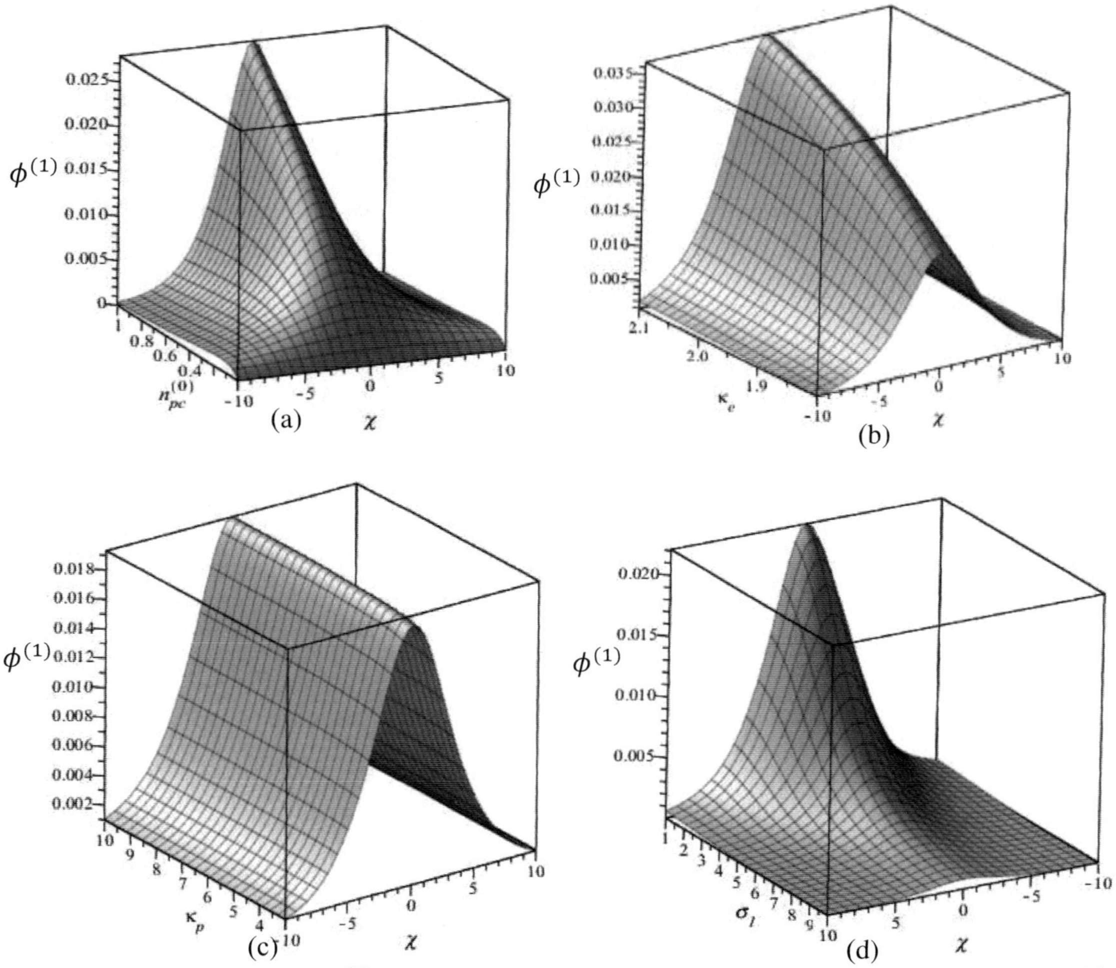

Figure 2. Effects of (a) $n_{pc}^{(0)}$ ($\kappa_e = 1.8, \kappa_p = 3$, $\mu_e = 0.2$, $\sigma_1 = 1$, $\mu_{ph} = 0.08$, $\sigma_2 = 0.5, u_{pc}^{(0)} = 0.6$, $V = 1.6$, $\phi^{(0)} = 0.1$, $U_0 = 0.02$) (b)κ_e with the same values of (a) but $n_{pc}^{(0)} = 0.5$, $\phi^{(0)} = 0.5$, and $\mu_e = 0.45$, (c)κ_p taking the same values of (a) but $\mu_{ph} = 0.12$, $n_{pc}^{(0)} = 0.5$ and $\mu_e = 0.45$, and (d) σ_1 with the same values of (b) but $n_{pc}^{(0)} = 0.71$, $\kappa_e = 1.8$ and $\mu_e = 0.5$ along with χ on the electrostatic PA mKdV solitary waves in weakly inhomogeneous plasmas.

5. Conclusions

The weakly unmagnetized inhomogeneous plasma system composing of immobile positive ions, mobile cold positrons, and superthermal hot positrons, as well as electrons, is considered. The KdV and mKdV equations with variable coefficients are derived using the RPM for investigating the nonlinear propagation of PASWs in the plasmas. It is found that the KdV equation admits both compressive and rarefactive solitons, but only compressive solitons are found from the mKdV equations in inhomogeneous plasmas. The contributions of density ratios, temperature ratios and superthermality parameters to the inhomogeneously solitons propagation characteristics are described. It can be concluded that the energy absorption appears by the solitons without changing the shape and velocity, thus the energy absorption enhances (suppresses) smoothly with the enhancement of plasma parameters for the rarefactive (compressive) PASWs in the plasma system. Furthermore, the system critically changes the polarity, which makes the energy absorption have maximum change near the critical values. The results obtained in this investigation might be useful for further verification in laboratory plasmas.

References

[1] Bremer J, Hoffmann P, Manson A H, Meek C E, Ruster R and Singer W 1996 *Ann. Geophys.* **14** 1317
[2] Franz J, Kintner P and Pickett J 1998 *Geophys. Res. Lett.* **25** 1277
[3] Pierrard V and Lemaire J 1996 *J. Geophys. Res.* **101** 7923
[4] Michel F C 1982 *Rev. Mod. Phys.* **54** 1
[5] Miller H R and Wiita P J 1987 *Active Galactic Nuclei* (Berlin: Springer) p. 202
[6] Shukla P K, Rao N N, Yu M Y and Tsintsadze N L 1986 *Phys. Rep.* **138** 1
[7] Berezhiani V, Tskhakaya D D and Shukla P K 1992 *Phys. Rev. A* **46** 6608
[8] Vasyliunas V M 1968 *J. Geophys. Res.* **73** 2839
[9] Baluku T K and Hellberg M A 2008 *Phys. Plasmas* **15** 123705
[10] El-Tantawy S A, El-Bedwehy N A and Moslem W M 2011 *Phys. Plasmas* **18** 052113
[11] Surko C M, Leventhal M and Passner A 1989 *Phys. Rev. Lett.* **62** 901
[12] Abdullah K, Haarsma L and Gabrielse G 1995 *Phys. Scr.* **T59** 337
[13] Shah M G, Hossen M R, Sultana S and Mamun A A 2015 *Chin. Phys. Lett.* **32** 085203
[14] Uddin M J, Alam M S and Mamun A A 2015 *Phys. Plasmas* **22** 062111
[15] Rahman M M, Alam M S and Mamun A A 2015 *Brazilian J. Phys.* **45** 314
[16] Rahman M M, Alam M S and Mamun A A 2015 *Astrophys. Space Sci.* **357** 36
[17] Shah M G, Hossen M R and Mamun A A 2015 *Brazilian J. Phys.* **45** 219
[18] Uddin M J, Alam M S and Mamun A A 2015 *Phys. Plasmas* **22** 022111
[19] Rahman M M, Alam M S and Mamun A A 2014 *J. Korean Physical Society* **64** 1828
[20] Alam M S, Uddin M J, Masud M M and Mamun A A 2014 *Chaos* **24** 033130
[21] Hafez M G and Talukder M R 2015 *Astrophys. Space Sci.* **359** 27
[22] Hafez M G, Talukder M R and Sakthivel R 2016 *Indian J. Phys.* **90** 603
[23] Hafez M G, Talukder M R and Ali M H 2017 *Plasma Phys. Reports* 43499
[24] Hafez M G, Talukder M R and Ali M H 2016 *Phys. Plasmas* **23** 012902
[25] Hafez M G, Talukder M R and Ali M H 2016 *Astrophy. Space Sci.* **361** 154
[26] Hafez M G, Roy N C, Talukder M R and Ali M H 2016 *Astrophys. Space Sci.* **361** 312
[27] Hafez M G, Roy N C, Talukder M R and Ali M H 2016 *Phys. Plasmas* **23** 082904
[28] Alam M S, Hafez M G, Talukder M R and Ali M H 2017 *Phys. Plasmas* **24** 072901
[29] Akter T, Deeba F and Kamal-Al-Hassan M 2016 *IEEE Transactions Plasma Sci.* **44** 1449
[30] El-Wakil S A, Zahran M A, El-Shewy E K and Mowafy A E 2006 *Phys. Scr.* **74** 503
[31] Mowafy A E, El-Shewy E K, Moslem W M and Zahran M A 2008 *Phys. Plasmas* **15** 708
[32] Attia M T, Zahran M A, El-Shewy E K and Mowafy A E 2010 *Z. Naturforsch. A* **65** 91
[33] Mowafy A E and Moslem W M 2012 *J. King Saud Univ.Sci.* **24** 343
[34] Gill T S, Kaur H and Saini N S 2006 *Pramana – J. Phys.* **66** 6
[35] Asano N 1974 *Suppl. Prog. Theor. Phys.* **55** 52

IOP Conf. Series: Journal of Physics: Conf. Series **1086** (2018) 012002 doi:10.1088/1742-6596/1086/1/012002

Effect of fluorine (F) on structural and electrical properties of sprayed ZnO thin films

A Rani[1], M K R Khan[1], M M Rahman[1], M S I Sarker[1] and M Kamruzzaman[2]

[1] Department of Physics, University of Rajshahi, Rajshahi-6205, Bangladesh
[2] Department of Physics, Begum Rokeya University, Rangpur-5400, Bangladesh

Email: anjali387.phy@gmail.com

Abstract. Un-doped and F doped ZnO thin films have been deposited on glass substrate using spray pyrolysis technique at substrate temperature 350 °C. The structural and electrical properties of these films were investigated by X-ray diffraction (XRD) and Van der Pauw method. Electrical transport study confirms semiconducting nature of the films. Dopant F atom enhances electrical conductivity substantially with the increase of F content. Activation energy measurement confirms that electrical transport mainly occurs due to free carrier transition and free excitation of defect states to the conduction band. The carrier concentration of ZnO increases with F-doping as confirmed by Hall voltage measurement and the order of carrier concentration is 10^{19} cm^{-3}.

1. Introduction

Zinc oxide (ZnO) is the most important semiconductor material suitable for microelectronic and shortwave optoelectronic applications owing to its high conductivity, chemical stability, nontoxicity and transparency, as well as being inexpensive and abundance in nature. ZnO is a type II-VI direct band gap (3.37 eV) semiconductor with high excitonic binding energy (60 meV at RT) exhibiting near UV emission [1-3]. Recently, ZnO) with hexagonal Wurzite crystal structure has been extensively investigated as an alternative material to ITO used in transparent electrodes, window materials in solar cells, light emitting diodes, flat panel displays, sensors etc. [4-8]. Zinc oxide thin films have attracted more attention due to their broad range of applications such as window layer in solar cells, piezoelectric nano-generators, photo detectors, surface acoustic wave device, gas sensors, biomedicine etc. [9-11].

Most of the researchers reported the effect of divalent and trivalent cation dopants on doped ZnO thin films. However, a few studies on F doped ZnO are found in the literature despite the fact that F can be an adequate anion doping candidate due to its similar ionic radius to the oxygen (F^{1-}- 0.136 nm, O^{2-}- 0.132 nm) [12, 13]. It is well known that in semiconductor, carriers can be originated from intrinsic donors and from extrinsic doping. In ZnO lattice intrinsic donors are oxygen vacancies present (if any) and Zn^{2+} atoms. On the other hand, extrinsic doping is substitution of F^{1-} for O^{2-} which will liberate one free electron for each F^{1-} substitution for O^{2-} and acts as a donor. Furthermore, when oxygen is replaced by fluorine in ZnO, it may passivate surface dangling bond resulting in reduction of carrier trapping. As a result, overall carrier concentration will be enhanced and contribute to electrical conduction. Therefore, F doped ZnO with high optical transparency can be used as an electrode or window layer for solar cell fabrication. Motivated with the fact described above, present study has been designed.

In the last few years, variety of deposition techniques have been employed to fabricate the ZnO thin films; including (RF) magnetron sputtering, sol gel, chemical vapor deposition, pulsed laser deposition and chemical spray pyrolysis etc. [14-19]. Among all these deposition techniques, the chemical spray pyrolysis (SP) is very simple, versatile and relatively cost-effective processing technique having the capability to produce large area, high quality adherent films of uniform thickness. The main focus of this work is to study the effect of F concentration on the microstructure and electrical transport properties of ZnO films and discussed.

2. Experimental

Un-doped and F doped ZnO films have been deposited on ordinary glass substrate by spray pyrolysis technique at substrate temperature 350 °C in air ambient. A homogeneous solution was prepared by dissolving zinc acetate dihydrate ($(CH_3.COO)_2Zn.2H_2O$) and ammonium fluoride (NH_4F) in distilled water at room temperature. The solution was sprayed through a commercially available medical infusion pipe onto the heated substrate by an air compressor. The glass substrates were cleaned by using liquid detergent, distilled water, piranha solution, ethanol and again distilled water. The cleaned substrate was placed on a substrate holder and heated by a simple electric heater. The temperature was controlled through a copper-constantan thermocouple. In preparation of all the filmsthe distance between tip of the spray nozzle and the surface of the substrate was kept at 20 cm, the rate of solution flow was 0.30 ml/min for half an hour and substrate temperature was 350 °C. The conditions for film deposition were kept constant so that thicknesses were same for all samples. The thicknesses of the films were (150 ±10) nm measured by Newton's ring method. Details of the film preparation by SP technique have been given elsewhere [20]. The chemical reactions that take place on the heated substrate are as follows:

$$(CH_3COO)_2Zn.2H_2O \xrightarrow[\text{Decomposes to}]{350^\circ C} ZnO + CO_2\,(g) + CH_4\,(g) + \text{Steam}\,(g)$$

$$(CH_3COO)_2Zn.2H_2O + NH_4F \xrightarrow[\text{Decomposes to}]{350\ ^\circ C} ZnO{:}F + CO_2\,(g) + NO\,(g) + CH_4\,(g) + \text{Steam}\,(g)$$

X-ray diffraction (XRD) study was performed using X-ray diffractometer, (D2 PHASER, BRUKER) with monochromatic CuK_α radiation of wavelength, λ = 0.1542 nm operated at accelerating voltage 30 kV and current 10 mA. X-ray diffraction patterns of the films were recorded by varying diffraction angle 2θ from 20° to 80°.Electrical resistivity and Hall voltage measurements were carried out by van der Pauw method [21].

3. Results and discussion

Figure 1 shows the XRD patterns of un-doped and F doped ZnO thin films deposited at 350 °C substrate temperature using spray pyrolysis technique indicating hexagonal Wurtzite crystal structure of the films. All the films have seven diffraction peaks corresponding to (100), (002), (101), (102), (110), (103) and (112) planes, which indicates the polycrystalline nature of ZnO films. No extra peaks other than ZnOwere observed. So, F atoms may be substituted O atoms or incorporated into interstitial sites in the ZnO lattice confirming the monophasic (ZnO) character of the grown films. The peak intensity of the films is different depending on the amount of F dopant. As seen from figure, the diffraction angle of all the peaks shift towards lower angle results in increase in a and c lattice parameters as given in table 1.

The crystallite size, ξ was estimated using Debye-Scherrer formula [22],

$$\xi = \frac{0.94\lambda}{\beta\cos\theta} \qquad (1)$$

where λ is the incident X-ray wavelength and β is the full-width at half maximum of the diffraction peak. The corresponding lattice parameters were calculated using the formula [23],

$$\frac{1}{d^2} = \frac{4}{3}\left\{\frac{h^2 + hk + l^2}{a^2}\right\} + \frac{l^2}{c^2} \qquad (2)$$

It is observed that for all the films lattice parameters *a* and *c* increased indicating the substitution of F atoms at the O sites, because of the difference of electronegativity between F and O ions. Each substitution of O atom by F atom gives one free electron in the system. However, observed intensity diminution in 2, 3 and 5% F doped samples may be due to incorporation of F atoms in the interstitial positions of host lattice.

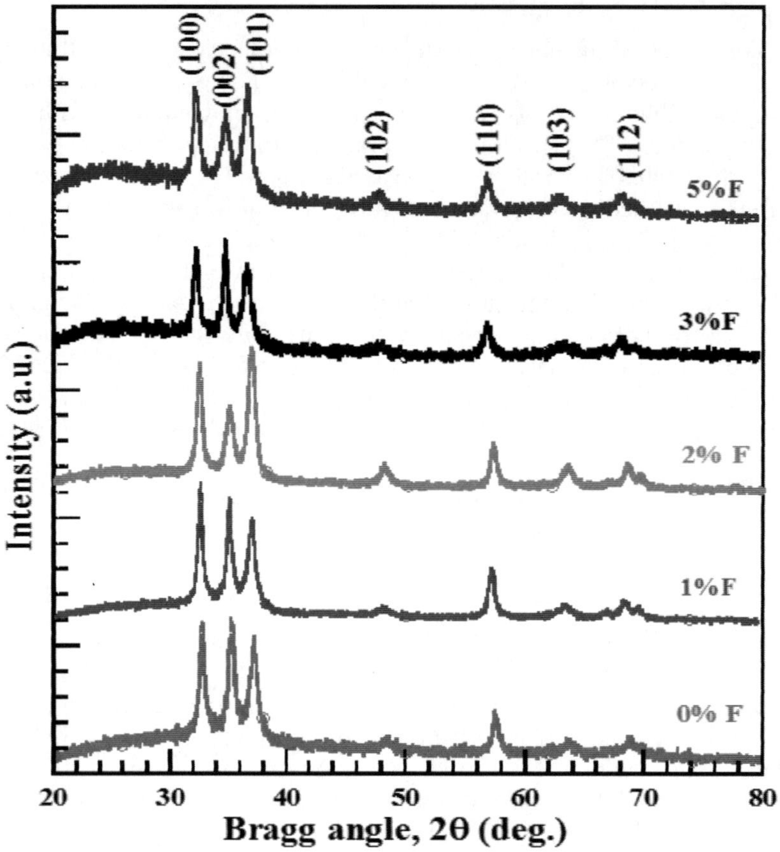

Figure 1. The XRD pattern of the un-doped and F doped ZnO thin films.

The lattice strain affects the crystal structures and properties of ZnO thin films to some extent and was calculated for (100) and (002) peaks using the formula,

$$\varepsilon = \frac{\beta}{4\tan\theta} \qquad (3)$$

The dislocation density represents the total length of dislocation lines per unit volume of the crystal. The dislocation density, δ for two major peaks (100) and (002) of all the samples were estimated using the Williamson and Smallmans relation [24],

$$\delta = \frac{1}{\xi^2} \qquad (4)$$

The calculated lattice strain and dislocation density of all the films are tabulated in table 1.

IOP Conf. Series: Journal of Physics: Conf. Series **1086** (2018) 012002 doi:10.1088/1742-6596/1086/1/012002

Table 1. Structural and defect parameters of un-doped and F doped ZnO films.

F conc. in ZnO (mol%)	Lattice parameter (nm)		Crystallite size, ξ (nm)		Lattice strain, $\varepsilon \times 10^{-3}$		Dislocation density, $\delta \times 10^{15}$ lines/m^2		*c/a* ratio
	a	*c*	(100)	(002)	(100)	(002)	(100)	(002)	
0	0.3154	0.5076	27	18	4.6714	6.4106	1.3717	3.0864	1.6089
1	0.3188	0.5131	55	20	6.4970	5.9933	3.3058	2.5000	1.6094
2	0.3199	0.5135	18	14	1.8668	2.2528	3.0864	5.1020	1.6052
3	0.3216	0.5171	44	36	2.9809	3.2990	0.5165	0.7716	1.6079
5	0.3220	0.5156	37	36	3.5831	0.7035	0.7035	0.7716	1.6012

Figure 2 shows the plot of electrical resistivity as a function of temperature for un-doped and F doped ZnO thin films. The resistivity found to decrease with the increase of temperature which confirms the semiconducting behaviour of ZnO films. This effect may be due to increase of transition of donor electrons to conduction band. It can also be seen that resistivity is strongly influenced by the doping and it decreases with F doping concentration. The RT resistivity for un-doped ZnO film is about 6.979 $\times 10^{-2}$ (Ω-cm) and it decreases to 5.939 $\times 10^{-2}$ (Ω-cm) for 5% F doped sample.

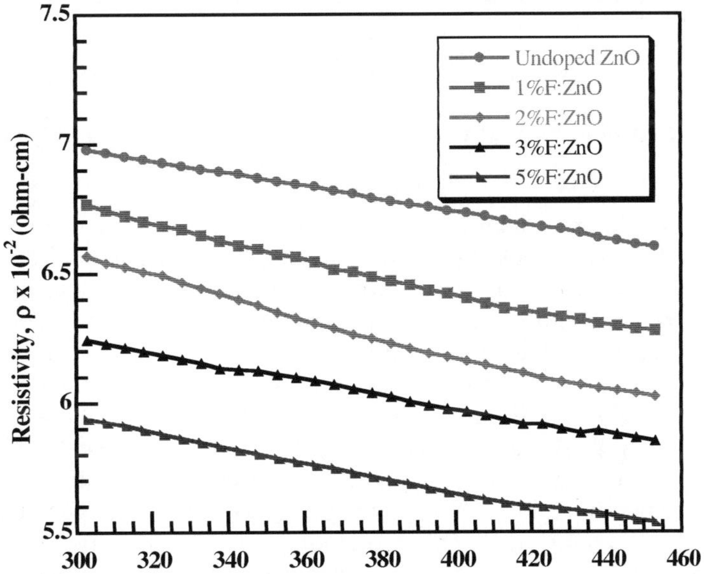

Figure 2. Variation of electrical resistivity with temperature.

Figure 3 illustrates the variation of RT resistivity (ρ), carrier concentration (n) and Hall mobility (μ_H) as a function of F concentration. As the F content increases, carrier concentration of the films increases while RT resistivity and mobility decrease.

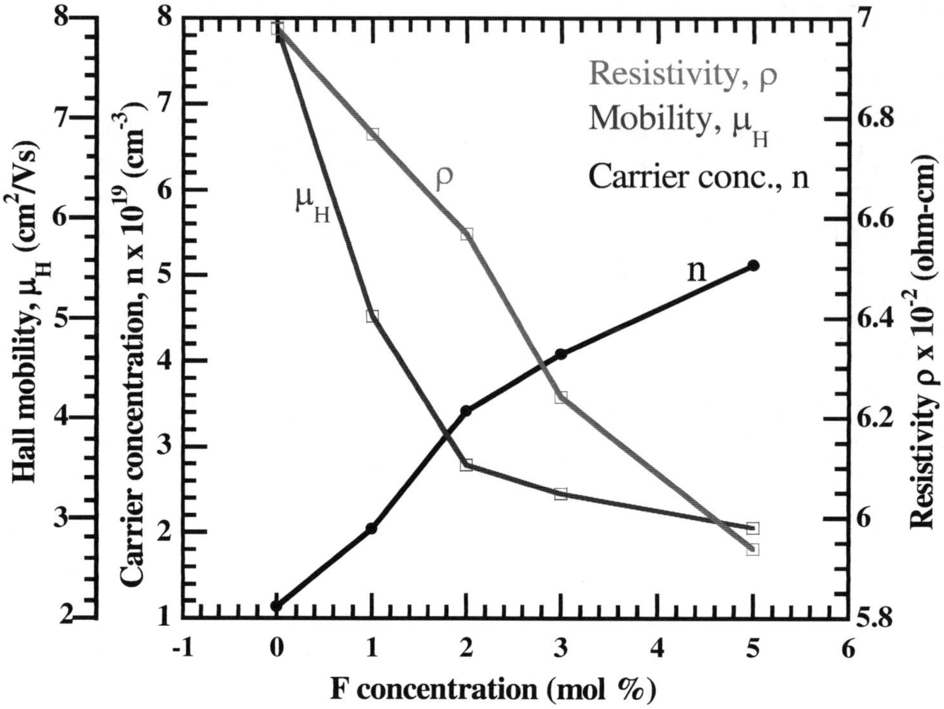

Figure 3. Variation in RT resistivity (ρ), carrier concentration (n) and Hall mobility (μ_H) as a function of F concentration.

The highest carrier concentration is about 1.2×10^{19} (cm^{-3}) for un-doped sample and it increases up to 5.1×10^{19} (cm^{-3}) for 5% F doped ZnO sample. On the other hand, highest mobility is found for un-doped sample. The resistivity of as deposited F doped ZnO films obtained in this study is found consistent with the resistivity of F doped ZnO film prepared by magnetron sputtering (as-grown) [25] and chemical spray [26] technique but higher than that of pulsed laser deposition method [27] as illustrated in table 2. The carrier concentration and mobility were also found to consistent [25]. However, after vacuum annealing of magnetron sputtered sample, they found lower resistivity, higher carrier concentration and mobility [25] as given in table 2. Wang et al [14] also found lower resistivity with higher carrier concentration and mobility by increasingsubstrate temperature from RT to 300 °C. In general, carrier mobility become lower if crystal grain size is small. Although, the crystallite size of un-doped and F doped films is found moderate in this study, however, mobility found to decrease with increasing F content. This is because, a part of free carriers may be trapped by surface dangling bonds and decrease mobility [28].

The activation energy ΔEcan be calculated using the relation,

$$\sigma = \sigma_0 \exp(-\frac{\Delta E}{2K_B T}) \qquad (5)$$

where, σ_0 is a constant and σ is the electrical conductivity, K_B is the Boltzmann constant and T is the absolute temperature. For the calculation of activation energy, the logarithmic variation of electrical conductivity with 1/T was taken for un-doped and F doped ZnO thin films as illustrated in figure 4. The activation energies of all the films were calculated from the slopes of the curves in the temperature range 303 K to 358 K (ΔE_1) and 358 K to 453 K (ΔE_2).

Table 2. RT resistivity (ρ), carrier concentration (n) and Hall mobility (μ_H) of un-doped and F doped ZnO films.

Preparation technique and conditions	F conc. in ZnO (mol%)	Resistivity, at (RT) $\rho \times 10^{-2}$ (Ω-cm)	Carrier concentration $n \times 10^{19}$ (cm^{-3})	Hall mobility, μ_H (cm^2/Vs)	Ref.
Spray 350 °C, air ambient	0	6.979	1.2	7.88	this work
	1	6.768	2.0	4.52	
	2	6.568	3.4	2.79	
	3	6.243	4.1	2.45	
	5	5.939	5.1	2.05	
Magnetron sputtering 150 °C	As-grown	7.84	1.7	4.7	[25]
	Vacuum annealed	0.36	5.6	30.5	
Chemical spray		6.83	-	-	[26]
Pulsed laser 300 °C		0.051	50.27	23	[27]

Thus more charge carriers overcome the activation energy barrier causing the electrical conduction. In the higher temperature region, the trapped charge carriers induce the electrical conduction at the grain boundaries with donor electron. Thus with the increase of temperature, the conduction mechanism changes from only thermally activated charge carriers to the grain boundary trapped charge conduction mechanism.The ΔE_1 and ΔE_2 values are tabulated in table 3.

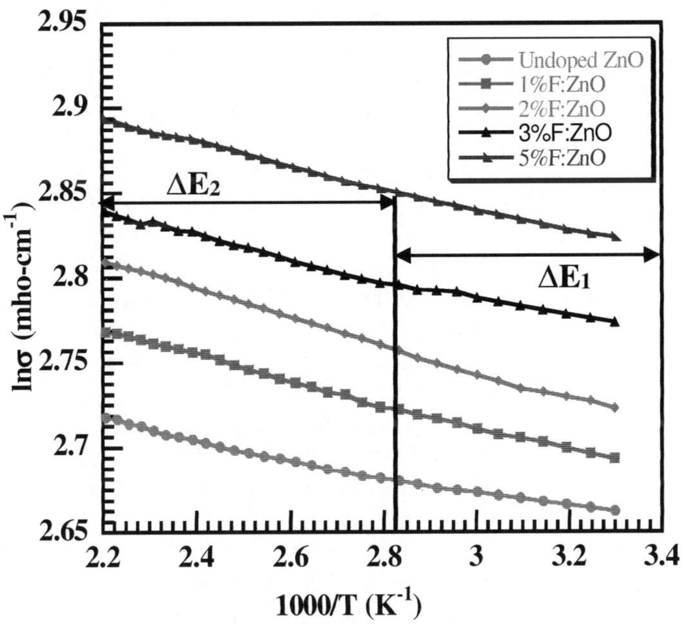

Figure 4. Variation of lnσ with 1000/T for un-doped and F doped ZnO thin films.

Table 3. Activation energy of un-doped and F doped ZnO thin films.

F conc. in ZnO (mol%)	Activation energy, ΔE (eV)	
	ΔE_1 (303-358) K	ΔE_2 (358-453) K
0	0.0115	0.0147
1	0.0105	0.0135
2	0.0062	0.0103
3	0.0089	0.0135
5	0.0095	0.0121

In the low temperature region, ΔE_1 varies from (0.0062 -0.0115) eV and in the high temperature region, ΔE_2 varies from (0.0103-0.0147) eV, though these variations are inconsistent with doping concentration of F. In the lower temperature region, as the temperature is increased, the electrons are excited thermally from donor levels to conduction band.

4. Conclusions

Un-doped and F doped ZnO thin films were deposited on glass substrate at 350 °C by spray pyrolysis technique. XRD characterization of all the films revealed polycrystalline nature. Hall measurement reveals un-doped and F doped ZnO films are n-type semiconductor and the effective carrier concentrations are found to be the order of ~10^{19} cm^{-3} and Hall mobility as high as 7.88 cm^2/Vs.

Acknowledgement

The authors are thankful to Centre of Super-Diamond and Advanced Films (COSDAF), City University of Hong Kong, China for providing the XRD facility.

References

[1] Shrisha B V, Bhat S, Kushavah D and Naik K G 2016 *Mater. Today. Proc.* **3** 1693
[2] Chen Y, Bagnal D M, Koh H J, Park K T, Hiraga K, Zhu Z Q and Yao T, 1998 *J. Appl. Phys.* **84** 3912
[3] Benkara S and Zerkout S, J. 2010 *Mater. Environ. Sci.* **1** 173
[4] Fortunato E, Barquinha P, Pimentel A, Goncalves A, Marques A, Pereira L and Martins R, 2005 *Thin Solid Films* **487** 205
[5] Iwata K et al. 2003 *Thin Solid Films* **445** 274
[6] Yoo J, Lee J, Kim S, Yoon K, Park I Jun, Dhungel S K, Karunagaran B, Mangalaraj D and Junsin Yi 2005 *Thin Solid Films* **480** 213
[7] Martı´nez M A, Herrero J and Gutie´rrez M T 1997 *Sol Energy Mater Sol Cells* **45** 75
[8] Wang Z L *J. NanosciNanotech* 2008. **8** 27
[9] Yoshida M M, Delgado F P, Lo´pez W E and Andrade E 2000 *Thin Solid Films* **376** 99
[10] Caglar M, Caglar Y and Ilican S 2007 *Phys. Status Solidi C***4** 1337
[11] Kerli S, Alver U, Yaykasli H, Avar B, Tanriverdi A and Kursun C 2013 *Asian Journal of Chemistry* **25** 7539
[12] Douayar A, Diaz R, Cherkaoui F and Moursli E, 2011 *Eur. Phys. J. Appl. Phys.***53** 20501.
[13] Tsai Y Z, Wang N F and Tsai C L 2010 *Thin Solid Films* **9** 4955
[14] Wang F H, Yang C F and Lee Y H 2014 *Nanoscale Research Letters* **9** 97
[15] Yoon H, Choi H and LeemJ-Y 2014 *Journal of the Korean Physical Society* **65** 509
[16] Keskenler E F, Turgut G, Aydin S and Dogan S 2012 *Scientific Research and Essays* **7** 3816
[17] Anusha M and Arivuoli D 2013 *Journal of Alloys and Compounds* **540** 131

[18] Cao L, Zhu L, Jiang J, Zhao R, Zhizhen Y and Zhao B 2011 *Solar Energy Materials & Solar Cell* **95** 894
[19] Castañeda L, Morales-Saavedra O G, Acosta D R, Maldonado A and Olvera M de la L 2006 *Phys. stat. sol.* **203** 1971
[20] Khan M K R**,** Rahman M A, Shahjahan M, Rahman M M, Hakim M A, Saha D K and Khan J U, 2010 *Current Appl. Phys.* **10** 790
[21] Van der Pauw L J 1958 *Philips Technical Rev*. (vol**20)** p 220
[22] Cullity B D 1972 *Elements of X-ray diffraction*(Addition-Wesley, Reading, MA) p 102
[23] Cullity B D and Stock S R 2001 *Elements of X-ray Diffraction* (New Jersey Prentice Hall) pp. 619
[24] Bilgin V, Kose S, Atay F and Akyuz I 2002 *Mater. Chem. Phys.***78** 51
[25] Yoon H S, Lee K S, Lee T S, Cheong B, Choi D K, Kim D H and Kim W M 2008 *SolarEnergy Materials and Solar Cells* **92** 1366
[26] Biswal R R, Velumani S, Babu B J, Maldonado A, Tirado-GuerraS, Castaneda L and Olvera de la L M 2010 *Materials Science and Engineering* B **174** 46
[27] Cao L, Zhu L P, Chen W F and Ye Z Z 2013 *Optical Materials* **35** 1293
[28] Liu B, Gu M, Liu X, Huang S and Ni C 2010 *Applied Phys. Letters* **97** 122101

IOP Conf. Series: Journal of Physics: Conf. Series **1086** (2018) 012003 doi:10.1088/1742-6596/1086/1/012003

Improvement of mechanical properties of polypropylene composite using filler, modifier and reinforcement

H Chowdhury, M M Rahman, M T Uddin and M M Rahman

Department of Chemical Engineering and Polymer Science, Shahjalal University of Science and Technology, Sylhet 3114, Bangladesh

Email: ifticep19@gmail.com

Abstract. This study focuses on improvement of mechanical and physical properties of polypropylene (PP) composite using filler, modifier and jute fiber as reinforcement. Rice husk ash (RHA), low-density polyethylene (LDPE) and jute fiber were used as filler, modifier, and reinforcement in the PP matrix, respectively. A series of test specimens were using various compositions of PP, fillers, modifiers, and reinforcement. Materials were mixed using two roll mixing machine, and specimens were prepared using injection molding machine. The specimens were evaluated on the basis of mechanical (tensile strength, elongation at break) and Physical (water absorption) properties. Effect of RHA was assessed on the basis of mechanical properties. Surface property of the PP composites was studied using Scanning Electron Microscope (SEM). The water absorption property of composites with various compositions was investigated. Further attempts had been made to optimize the three process parameters (RHA, LDPE, jute fiber) in the PP composite with respect to its tensile strength. The optimum parameters for the PP composite were found to be 10 wt% RHA, 10 wt% LDPE and 25 wt% jute reinforcement, respectively in the PP matrix.

1. Introduction

Technological development mostly depends on advancements in the field of engineering materials. Composite materials in this regard represent nothing less than a giant step in the ever constant effort toward optimization in materials. Over past several years, there has been an increasing interest in the use of polymer composite due to their various advantages over other materials. This lightweight type of composites allows for lower fuel consumption which can be used in a broad range of applications, including in the aerospace, automotive, and rail sectors [1].

The use of lignocellulosic fibers as the reinforced materials in thermoplastics has received a lot of interest due to their low densities, low cost, and non-abrasive nature [2], which minimizes the wear of processing equipment and so forth [3-4]. Compared with thermosets, thermoplastic composites materials typically have higher strain to failure, faster to consolidate and retain their ability to be recycled [5]. Among other thermoplastic polymers, Polypropylene (PP) provides most of the advantages with regards to economic (price), ecological (recycling behavior) and other technical requirements [6]. Previously, jute fiber was used as a reinforcing agent in the polymer matrix with various advantages and minimal environmental effect because of its biodegradable properties [7, 8]. Commercially available silica fillers significantly improve composite mechanical properties [9, 10].

In Bangladesh, huge quantities of rice husk ash (RHA) have been generated as waste materials from the rice processing mill. Depending on the combustion conditions, RHA contains a large quantity of silica which is in partly crystalline and amorphous forms, the rest being an amalgamation of carbon

IOP Conf. Series: Journal of Physics: Conf. Series **1086** (2018) 012003 doi:10.1088/1742-6596/1086/1/012003

impurity and a small fraction of metal oxide impurity. The composition of the rice husk ash is given below [11].

Table 1. The composition of the rice husk ash

Content, wt.%							
SiO_2	Al_2O_3	Fe_2O_3	CaO	MgO	K_2O	Na_2O	P_2O_5
93.4	0.05	0.06	0.31	0.35	1.4	0.1	0.8

In this study, the technical feasibility of RHA as filler materials in the PP matrix has been studied. Effect of RHA, LDPE and jute fiber in the PP matrix has been evaluated separately, and optimum content of the filler, modifier, and reinforcement have been determined. Micro-structural analysis (SEM) and water absorption characteristics of different PP composites have also been reported.

2. Materials and Methods

2.1. Materials

Materials used in this study are Polypropylene (PP) pellets as matrix material, jute fiber as reinforcement and LDPE as modifier and RHA as filler. The PP was supplied by Polyolefin Company Pvt., Ltd., Singapore. The PP has a density of 0.905 g/cm^3 at room temperature and the Melting Temperature (T_m) and Melt Flow Index are 160°C and 6 g/10 min., respectively. The jute fiber was collected from local market of Gazipur, Dhaka, Bangladesh. LDPE was commercial grade, granular and virgin. It was manufactured by The Polyolefin Company Pvt., Ltd., Singapore. RHA was used as filler in the composite which was bought from local market of Gazipur, Dhaka, Bangladesh.

2.2. Jute preparation

The jute fiber was first washed and then dried in an oven at 40°C for 24h. The dried jute fiber was then cut into 2-3 mm in length for the fabrication of jute-PP composites.

2.3. Rice husk ash (RHA) preparation

Rice husk ash collected from local rice processing station. The blackish rice husk contained carbonaceous material. In view to burn carbonaceous material completely, it was heated at 840°C for an hour in the Muffle furnace. The ash was then collected from the furnace. It was turned into a grayish color which was used as filler in the PP matrix.

2.4. Composite fabrication

In this study, three different types of composite specimens were prepared.

i. RHA blended with PP to prepare RHA-PP composite
ii. Jute fiber reinforced PP composite
iii. LDPE mixed with PP to prepare LDPE-PP composite

PP and the other ingredient was mixed using two roller mixing machine by applying heat at 130-140°C to turn it into a uniform mass. This mixing results in a sheet of PP with other ingredients. The prepared sheet was then cut into small pieces and fed into injection molding machine. In the injection chamber, the molding compound is heated at a temperature about 110°C and therefore it changes into fluid form and after that forced through an orifice into the desired dog bone shaped mold which is kept relatively cool. The composite solidifies in the mold from which it can then be removed for further testing.

2.5. Evaluation of specimens

Mechanical (tensile strength, elongation at break) and physical (water absorption) properties of the polymer composites was evaluated in the study.

Tensile test

The tensile strength of the PP composite specimens was measured following ASTM D638 01 [12] using a Universal Testing Machine (Model DRK101-300; Shandong Drick Instruments Co. Ltd, China) at a cross-head speed of 25mm/min. The size of the sample used for tensile strength test was

150 mm × 10 mm × 3.2 mm. Five specimens of each type of composite were tested and the average values of Tensile Strength and %Elongation were reported by calculating of maximum four values.

Water Absorption

Specimens of the composites were immersed in the static water glass beaker at room temperature for different time periods (up to 72 hours). Before immersion, the weight of the samples was taken. At certain time intervals, samples were taken out from the beaker and wiped smoothly with tissue paper and after that weighed promptly. Percent of water uptake was then calculated using the following equation:

$$W_{up}(\%) = \{(W_f - W_i) / W_i\} \times 100\% \qquad (1)$$

Where, W_{up} = water absorption, wt%

W_i = weight of the sample before immersion in water

W_f = weight of the sample after immersion

2.6. Micro-structural analysis

Micro-structural analysis of PP composites with varying compositions was conducted using Scanning Electron Microscope (SEM, Model: JSM-6701F) supplied by JEOL Company Limited, Japan.

3. Results and Discussion

3.1. Tensile strength

Effect of RHA content on the tensile strength is presented in Figure 1. The figure shows that addition of RHA up to 10 wt% in the PP matrix improves the tensile strength; further addition of RHA reduces the strength. The initial increase in strength of the composite is due to good filler-matrix interactions, which also enable more stress to be transferred from the matrix to the fillers during external loading. Above 10 wt% of RHA, with further increase in filler content, the matrix-filler aggregates deform easily in the composite due to the reduction of binder matrix (PP) and an increase of filler material, which tend to reduce the strength of the composite.

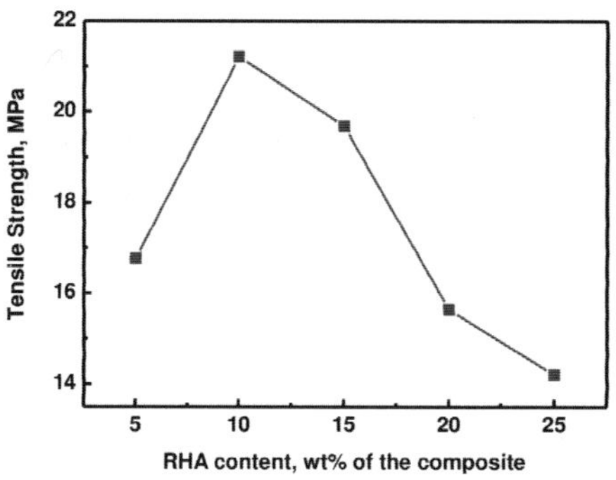

Figure 1. Tensile strength as a function of RHA content, wt%

Variation in the tensile strength of PP composite with jute fiber content is presented in Figure 2. The Figure shows that the tensile strength of jute-PP composite increases up to 25 wt% jute fiber content in the PP matrix, thereafter strength decreases with the further addition of jute fiber content in the matrix. The initial increment in strength can be explained by the interfacial bond formation between polymer and jute. After the optimum composition of PP matrix with 25 wt% jute fiber loading, the figure shows a reduction in strength because of improper mixing and the weak interfacial area between the fiber and matrix. A large content of jute fiber prevents the polymer melt from wetting the jute fiber during the molding.

IOP Conf. Series: Journal of Physics: Conf. Series **1086** (2018) 012003 doi:10.1088/1742-6596/1086/1/012003

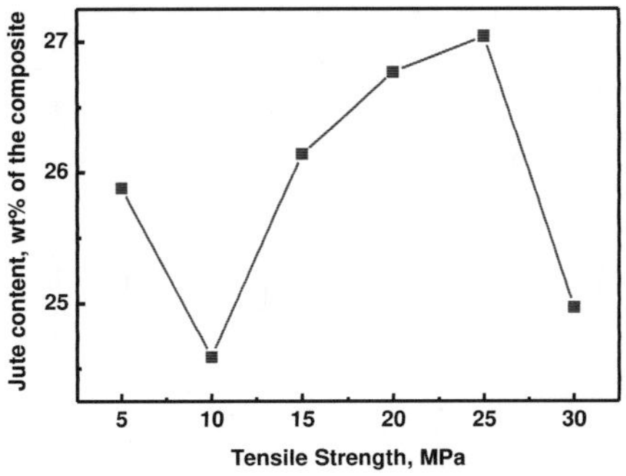

Figure 2. Tensile strength as a function of jute content, wt%

Blends of PP and LDPE contribute to making recycling more economically attractive. Effect of LDPE content on the tensile strength of PP composite is presented in Figure 3. The figure shows that PP composite containing 10 wt% of LDPE shows optimum tensile strength. The effect of LDPE in the PP matrix is as similar as RHA. Addition of LDPE up to 10 wt% improves the tensile strength; further incorporation of LDPE in the PP matrix reduces the strength. At lower filler content, matrix-filler interactions are quite good and the tensile strength is better in that region. After the optimum point, amount of binder matrix material continually decreases as the LDPE content increases which is the reason behind matrix-filler deformation and lower tensile properties.

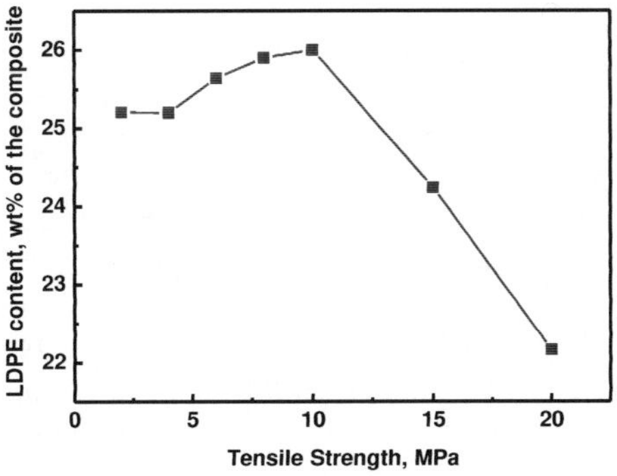

Figure 3. Tensile strength as a function of LDPE content, wt%

3.2. Elongation at break

Figure 4 shows variation in elongation of PP-RHA composite with RHA content in the composition. The figure indicates that addition of RHA in the PP matrix decreases elongation of the PP composite. As tensile strength increases with increasing filler content, the elongation decreases simultaneously and vice versa.

IOP Conf. Series: Journal of Physics: Conf. Series **1086** (2018) 012003 doi:10.1088/1742-6596/1086/1/012003

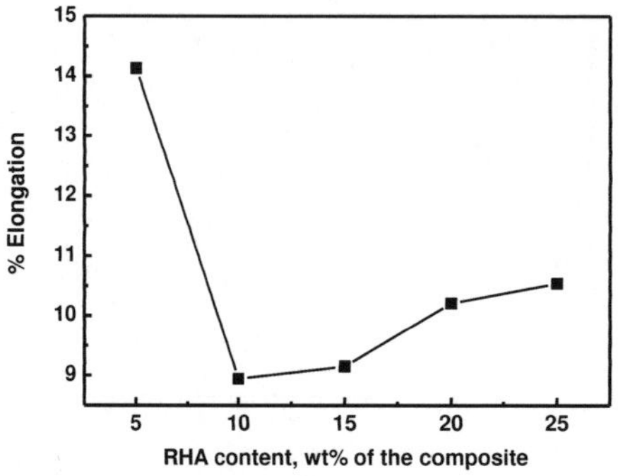

Figure 4. %Elongation as a function of RHA content, wt%

Variation in elongation behavior with jute fiber content in the PP matrix is presented in Figure 5. The figure shows that elongation decreases with the increased jute fiber content in the PP matrix. The cause behind the decrease is the cross-linking between the neighboring jute with Polypropylene chain which increases the rigidity and toughness of the composite material but decreases its ductility.

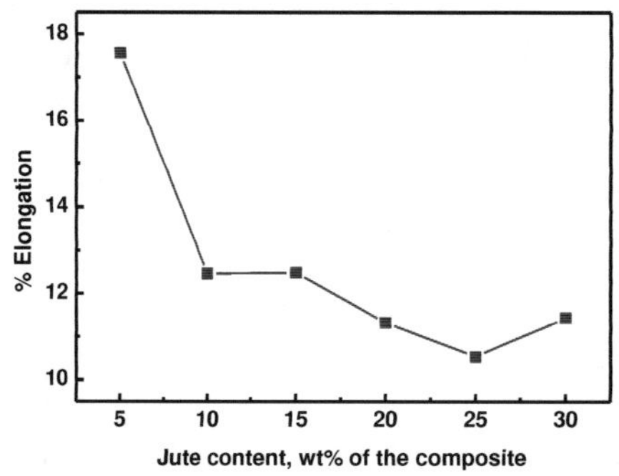

Figure 5. %Elongation as a function of jute content, wt%

The elongation at break for the different content of LDPE in the PP matrix is presented in Figure 6. It is found that PP composite with 10 wt% LDPE shows higher elongation at break.

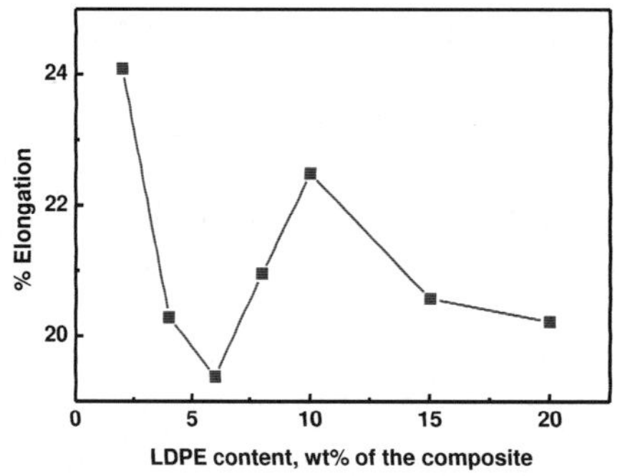

Figure 6. %Elongation as a function of LDPE content, wt%

3.3. Water absorption

The water absorption capacity of PP composites with different RHA content is presented in the Fig. 7. The figure shows that water absorption increases with increasing filler RHA content. This result is expected since RHA is hydrophilic in nature, the amount of water uptake increases significantly with increasing RHA content.

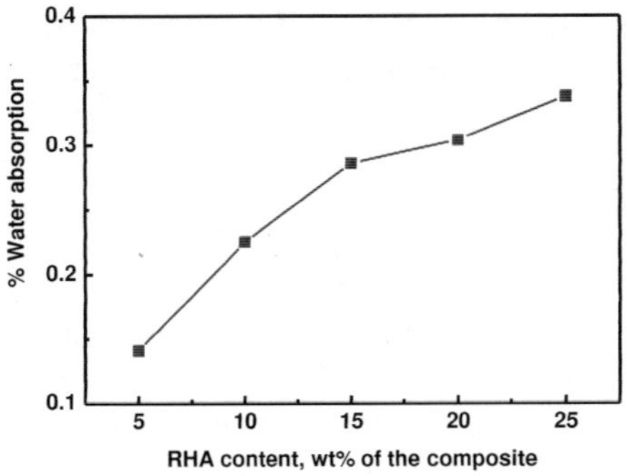

Figure 7. Water absorption capacity as a function of RHA content, wt%

Figure 8 shows the water absorption capacity of PP composite as a function of jute fiber content. The figure shows that water absorption capacity increases with increasing jute fiber contents in the matrix. The reason is that the PP matrix is hydrophobic and a layer of PP always repels water to penetrate it into the composite. Since the jute is highly hydrophilic in nature and as the jute content increase in the composite, some of the jute pulls out from the PP matrix and absorb water and this leads to the further uptake of water into the composite matrix.

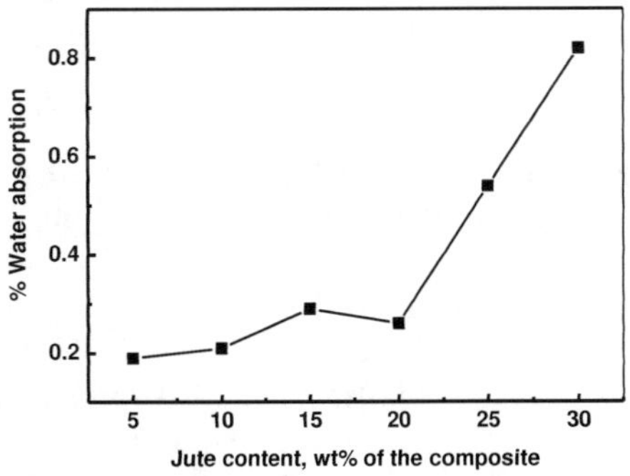

Figure 8. Water absorption capacity as a function of jute content, wt%

3.4. Micro-structural analysis

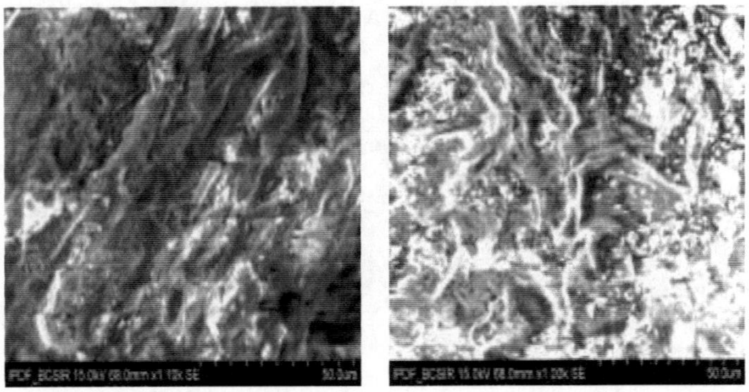

(a) (b)

Figure 9. SEM micrograph (a) PP with 10 wt% $CaCO_3$ and (b) PP with 10 wt% RHA

Micro-structure of PP-$CaCO_3$ and PP-RHA composite have been analyzed using SEM. Figure 9 clearly indicates improved crystallinity in the RHA-PP composite than the $CaCO_3$-PP composite. So, RHA can be a good replacement of conventional $CaCO_3$ filler in the composite without degrading the composite mechanical properties.

4. Conclusions

Rice husk ash was used as alternative fillers in replace of conventional filler calcium carbonate in the PP matrix. The tensile strength and elongation at break of the RHA blended PP composites was found maximum at 10 wt% RHA content in the composite. The SEM micrograph of RHA-PP composite showed improved crystallinity than $CaCO_3$-PP composite. Therefore, RHA can be used as an alternative of $CaCO_3$ in PP composite which is environment friendly, biodegradable and cheap filler. Water absorption properties of the different specimen showed that water absorption capacity increases with both RHA and jute content in the PP matrix. Based on mechanical properties, the optimum parameters for the PP composite was found to be 10 wt% RHA, 10 wt% LDPE and 25 wt% jute reinforcement, respectively and in the PP matrix.

Acknowledgment

The authors would like to acknowledge financial support from Shahjalal University of Science and Technology (SUST) Research Grant 2016-2017 (Project code: AS-04).

References

[1] Chawla K K and Rao A P 1998 *Composite Materials Science and Engineering* **2** 3-4
[2] Woodhams R T, Thomas G A and Rodgers D K 1984 *Polymer Engineering Science* **24** 1166-71
[3] Zadorecki P A and Michell A J 1990 *Composites Science and Technology* **10** 69-77
[4] Bismarck A V, Askargorta I A and Springer J P 2002 *Polymer Composites* **23** 872-94
[5] Vilaseca F H, Mendez J V, Pelach A P and Canigueral P N 2007 *Process Biochemistry* **42** 329-34
[6] Planckett D K, Andersen T L, Pedersen W B, and Nielsen L V 2003 *Composites Science and Technology* **63** 1287-96
[7] Palau T H, Garcia F J, Barcelo O and Martinez M 2001 *International Journal of Adhesion and Adhesives* **21** 1–9
[8] Wang M J, Mahmud A K, Murphy U J and Patterson W J 1998 *Kautschuk und Gummi Kunststoffe* **51** 348-60
[9] Gassan J A and Bledzki W K 1999 *Composites Science and Technology* **59** 1303-09
[10] Sarkar D A, Das B K and Ray A J 2004 *Composites Science and Technology* **64** 2213-19
[11] Ray A J, Sarkar D A, Das B K and Rana A K 2002 *Composites Science and Technology* **62** 911-17
[12] ASTM Standard D 638-01, *Annual Book of ASTM Standard* **8** ch 3 (2002).

Microstructural, morphological and electrical properties of sol-gel derived $CoFe_2O_4$ nanomaterials

A Hossain, M S I Sarker, M K R Khan and M M Rahman

Department of Physics, University of Rajshahi, Rajshahi-6205, Bangladesh.

E-mail: haliru10@gmail.com

Abstract. $CoFe_2O_4$ nanomaterials have been synthesized through conventional sol-gel process by using $CoCl_2.6H_2O$ and $FeCl_3$ as precursors and citric acid is used as a capping agent. The as synthesized nanomaterials have been investigated thoroughly using x-ray diffraction (XRD) technique, scanning electron microscopy (SEM) and an impedance analyzer. The XRD patterns indicate the formation of body centered cubic spinel $CoFe_2O_4$ having cell constant 8.403 Å and the corresponding space group is *Fd-3m*, in which the Fe^{3+} and Co^{2+} ions occupied in octahedral and tetrahedral sites in the crystallographic orientations. The average crystallite size was found to be 30 nm. The SEM micrographs reveal that the synthesized nanomaterials (NMs) formed as octahedron and tetrahedron and the particles are well dispersed and contain some pores. We found 17.5% porosity of the nanomaterials that plays an important role in technological applications, e.g. water and chemical filtration related to bio-medical applications. Measurement of the AC electrical properties shows that dielectric constant increases up to 5.13 for 0.3 MHz and then it decreases with increasing applied frequency. Lower dielectric constant at high frequency region appears due to decreasing interfacial polarization and limiting grain boundary effect.

1. Introduction

Generally, magnetic metals and metal oxides have a lot of attraction in materials science and engineering due to their wide range of applicability and outstanding properties. The cobalt ferrite nanomaterials (NMs) are used in various applications and have many efforts because of their unique physical and chemical properties such as low dielectric constant, high electrical resistance as well as excellent chemical stability and mechanical hardness [1-4]. The cobalt ferrite NMs has cubic spinel structure and its properties are affected by composition and microstructure cause of octahedral and tetrahedral occupancy in the lattice sites. For these properties cobalt ferrite is technologically important and suitable for high density recording media [2-3], Ferro fluid technology, biomedicine, magnetic resonance imaging, biosensors, magnetic hyperthermia therapy [4-5], data storage, magnetic refrigerators and microwave devices. These applications are strongly affected by shape and size of the NMs, hence the control of the shape and size of the NMs are quite important. Different types of chemical routes have been used to synthesize the $CoFe_2O_4$ NMs such as sol-gel method [6-8], micro emulsion, co-precipitation method [9], polyol reduction method [10], laser ablation, Turkevitch method and transmetallation technique [9] etc. Among these methods, we used sol-gel method for the synthesis of $CoFe_2O_4$ NMs because it is cost-effective and low temperature operation and an excellent chemical route for obtaining fine nanocrystalline and homogeneous powder.

In this work, to synthesis $CoFe_2O_4$ NMs through sol-gel route, ferric chloride and cobalt chloride are used as a precursors and citric acid as a capping agent. We have studied the effect of annealing temperature on its structural properties through XRD technique. The surface morphologies, chemical composition and AC electrical properties have also been examined for sample annealed at 600 °C.

2. Experimental Procedure

In order to synthesize the $CoFe_2O_4$ NMs using a citrate assisted sol-gel route stoichiometric amounts of 1.19 g $CoCl_2$. $6H_2O$ (97%: Merck), and 0.82 g $FeCl_3$ (98%: Merck) dissolved in 200 ml distilled water and a few drops of concentrated nitric acid were also added to prevent hydrolysis. Then, 8.46 g citric acid and a few ml of dimethylamine were also added to prevent the agglomeration of the particles. Thereafter, the mixtures of the solution were magnetically stirred at 60 °C for 2 hours and then evaporated at 100 °C to form a gel. The gel mixtures were dried at 150 °C for 18 hours and the resulting powder was decomposed in air and heated at 250 °C for 3 hours inside a furnace at air ambient. The powder samples were ground thoroughly about 2 hours using an agate mortar pestle and heat treated further at different temperatures for 4 hours at air ambient. Structural features such as crystallite size and crystallographic structure were studied for different annealed temperatures of the as synthesized NMs using Rigaku Smart Lab X-ray diffractometer using Cu-Kα radiation of λ = 1.54184 Å. The surface morphological properties were examined through a field emission scanning electron microscopy (FE-SEM) with accelerating voltage 5 keV (Philips XL30 FEG). The atoms present in the NMs composition and the ratio of the atoms were confirmed by energy dispersive x-ray (EDAX) intensity. The AC electrical measurements were obtained using an impedance analyser (Agilent technologies, model: 4294A) with variation of frequency from 50 kHz to 1 MHz at room temperature. A disk shaped pellet of 13.1 mm in diameter and 2.9 mm in thickness was prepared to determine electrical properties. The surfaces of the disk were roughly polished and then placed in the specimen holder for the measurements.

3. Results and Discussion

3.1. Structural Characterization

Figure 1 shows the XRD patterns of the as synthesized $CoFe_2O_4$ NMs annealed in air at temperatures 400 °C, 600 °C, 700 °C, 800 °C and 900 °C, respectively. It is seen from figure 1 that the observed peaks (220), (311), (222), (400), (422), (511) and (440) are the indication of spinel $CoFe_2O_4$ body centred cubic structure with space group *Fd-3m* (227), having lattice parameter '*a*' is equal to 8.403 Å (JCPDS card: 22-1086) [2, 6-8]. In the spinel structure, the cations Co^{2+} and Fe^{3+} may be occupied into two different sites, i.e. octahedral and tetrahedral sites within the lattice structure [2, 10]. The average crystallite size was determined from XRD peak broadening using the Debye-Scherrer formula [12], $D = K_\alpha \lambda / \beta \cos\theta$, where, λ is the wavelength of the x-ray radiation, K_α is a constant (0.94), β is the full width at half maximum in radian, θ is the diffracting angle. The average crystallite size was found to be 30 nm. Besides, the peaks around $2\theta = 44°$ and 65° indicate trace amount of FeCo alloy formation in the composition [9-10, 13]. The FeCo phases were observed with body centred cubic structure with space group *Pm-3m*. The lattice parameter '*a*' is equal to 2.954 Å. Moreover, at 400 °C the impurity phases of hematite (Fe_2O_3) and cobalt oxide (Co_3O_4) were confirmed from XRD patterns, which also reduces at 600 °C [6, 9-10]. The average crystallite size becomes larger with increasing annealing temperatures and the impurity phases were also reproduced due to decomposed of NMs. Thus, the optimal temperature to synthesize the $CoFe_2O_4$ NMs is 600 °C.

IOP Conf. Series: Journal of Physics: Conf. Series **1086** (2018) 012004 doi:10.1088/1742-6596/1086/1/012004

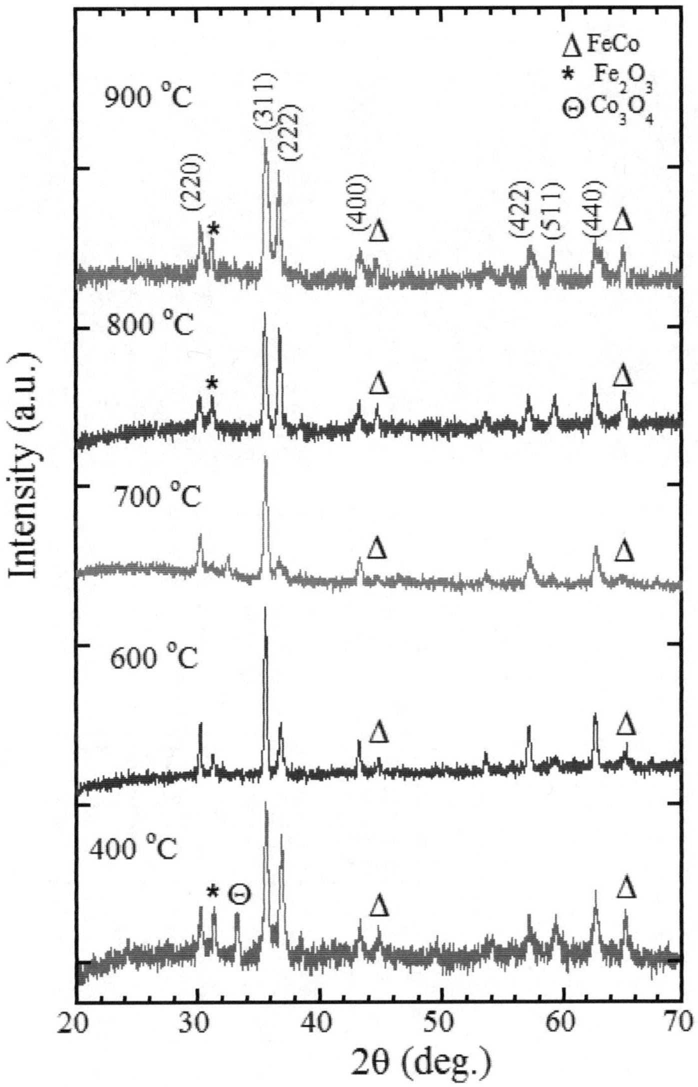

Figure 1. X-ray diffraction patterns of the as synthesized $CoFe_2O_4$ nanomaterials.

3.2. Surface Morphologies

Figure 2 shows the SEM micrographs of the as synthesized cobalt ferrite NMs for samples annealed at 600 °C. It is seen from the SEM micrographs in figure 2 that the synthesized NMs are formed as octahedron and tetrahedron and the particles are well dispersed and the mean particle size is about 30 nm [7]. The microstructure of the synthesized NMs reveals the grains are perfectly ordered with porous structure. The porosity was determined using the formula, $P = 1 - (\rho_m/\rho_x)$, where, ρ_m and ρ_x are measured density and x-ray density of the NMs, respectively [14]. The porosity of the NMs was found to be 17.5%.

IOP Conf. Series: Journal of Physics: Conf. Series **1086** (2018) 012004 doi:10.1088/1742-6596/1086/1/012004

Figure 2. The SEM micrographs of the as synthesized $CoFe_2O_4$ NM samples annealed at 600 °C.

The compositional analysis of the $CoFe_2O_4$ NMs was performed through EDAX analysis. It is seen from figure 3 that the ratio of Fe and Co atoms are 49.37% and 50.63%, respectively. It is important to notice that the EDAX analysis shows the ratio of Fe and Co is very close to 50:50 (1:1) which is in good agreement with nominal composition.

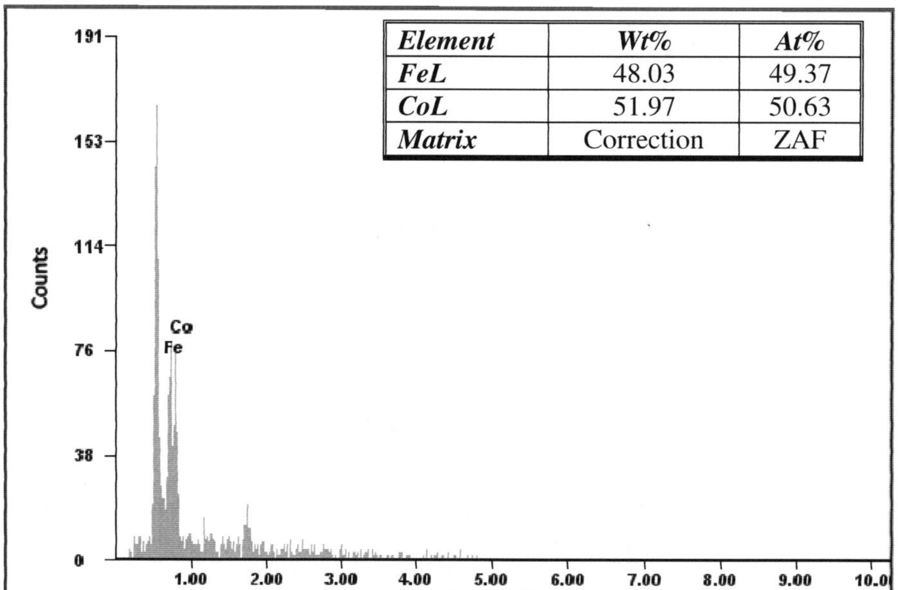

Element	*Wt%*	*At%*
FeL	48.03	49.37
CoL	51.97	50.63
Matrix	Correction	ZAF

Figure 3. The EDAX spectra of the as synthesized $CoFe_2O_4$ nanomaterials.

3.3. AC Electrical Properties

The AC electrical properties were measured using an impedance analyser at room temperature. Figure 4 shows the dielectric constant, k and capacitance, C as a function of frequency for the cobalt ferrite NMs. The dielectric properties of materials are affected by synthesis technique, temperature, percentage of composition, density, crystal structure, porosity and heat treatment, etc. In an ac fields,

IOP Conf. Series: Journal of Physics: Conf. Series **1086** (2018) 012004 doi:10.1088/1742-6596/1086/1/012004

dielectric constant is a complex quantity and we measured the real part of this quantity as a function of frequency. The dielectric constant is increased steeply with the increasing of frequency up to 0.2 MHz and holds maximum value 5.13 at 0.3 MHz, and then it decreases with increasing applied frequency. This results indicate the dielectric relaxation at low frequency to dispersion phenomenon at high frequency [2].

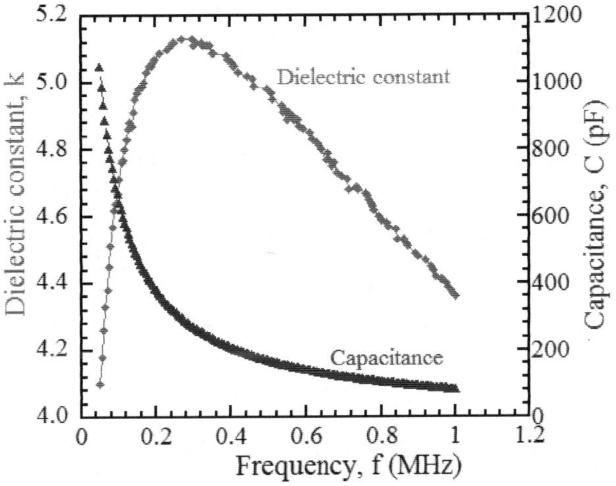

Figure 4. The dielectric constant and capacitance of the as synthesized samples annealed at 600 °C.

According to Maxwell–Wagner model, the grains are separated by poorly conducting grain boundaries [15]. At first, the dielectric constant increases due to increasing polarization with applied frequency. After certain high frequency, the electron cannot keep up with the applied field, i.e. electrons change their direction that decreases the probability of electrons to reach the grain boundaries. The grain boundaries weaken at higher frequencies. As a result, the polarization decreases and hence also the dielectric constant. Besides, from figure 4, it is seen that the capacitance of the cobalt ferrite NMs decreases with increasing frequency due to diminishing grain boundary effect [2, 13].
Figure 5(a) illustrates the ac conductivity and ac resistance of the cobalt ferrite NMs and 5(b)

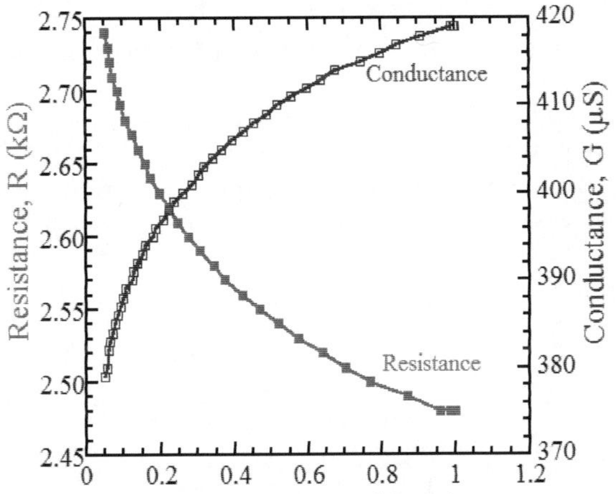

Figure 5(a). Conductance and resistance of the as synthesized materials for sample annealed at 600 °C.

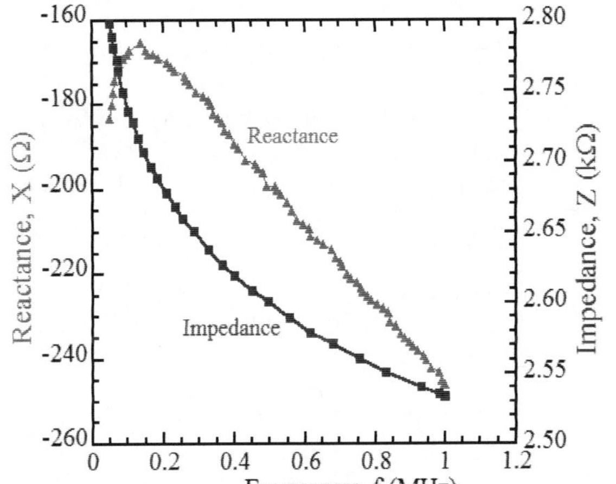

Figure 5(b). Reactance and impedance of the as synthesized materials for sample annealed at 600 °C.

represents the reactance and impedance of the synthesized NMs with applied frequency. The ac conductivity of dielectric materials is influenced by the local electric charge as well as charge hopping process [16]. It is observed from figure 5(a) that the conductivity increases with increasing frequency due to the increasing of the mobility of the electron. From figure 5(b) it is noticed that the behavioursof the impedance are very similar to AC resistance as shown in figure 5(a), which decreases with increasing applied frequency due to decreasing interfacial polarization. The reactance gradually increased with the increasing of frequency, after certain maximum value it falls down due to the limiting grain boundaries of the NMs.

4. Conclusions

The synthesis, microstructural, surface morphological and frequency dependent electrical properties of $CoFe_2O_4$ NMs using sol-gel route is explained. The XRD patterns confirm the crystalline phase of the synthesis NMs is spinel $CoFe_2O_4$. The SEM micrographs show the $CoFe_2O_4$ NMs formed as octahedron and tetrahedron. The synthesized $CoFe_2O_4$ NMs contains some pores, that play an important role in water and chemical filtration related to bio-medical applications. The high resistance and low dielectric constant indicate the low eddy loss of current, which make the NMs as a promising candidate for high frequency components.

References

[1] Rajath P C, Manna R S, Banerjee D, Varma M R, Suresh K G and Nigam A K 2008 *J. Alloys and Comp.* **453** 298
[2] Chen W, Zhu W, Tan O K and Chen X F 2010 *J. Appl. Phys.* **108** 034101
[3] Khandekar M S, Kamble R C, Patil J Y, Kolekar Y D and Suryavanshi S S 2011 *J. Alloys and Comp.* **509** 1861
[4] Deraz N M 2010 *J. Analytical and Applied Pyrolysis* **88** 103
[5] Kumar L, Kumar P, Narayan A and Kar M 2013 *International Nano Letters* **3** 8
[6] Nautial P, Seikh M M, Lebedev O I and Kundu A K 2015 *J. Magn. Magn. Mater.***377** 402
[7] Sinkó K, Manek E, Meiszterics A, Havancsák K, Vainio U and Peterlik H 2012 *J. Nanopart. Res.***14** 894
[8] Rao K S, Choudary G S V R K, Rao K H and Sujatha C H 2015 *Procedia Materials Science* **10** 19
[9] Thirumal E, Prabhu D, Chattopadhyay K and Ravichandran V 2010 *Phys. Status Solidi* A, **207** 2505
[10] Zehani K, Bez R, Boutahar A, Hlil E K, Lassri H, Moscovici J, Mliki N and Bessais L 2014 *J. of Alloys and Comp.* **591** 58
[11] Shadrokh S, Farahmandjou M and Firozabadi T P 2016 *Phys. Chem. Res.***4** 153
[12] Scherrer P 1918 *Göttingen Nachrichten Gesell* **2** 98
[13] Nongjai R, Khan S, Asokan K, Ahmed H and Khan I 2012 *J. Appl. Phys.* **112** 084321
[14] Kumar H, Srivastava R C, Negi P, Agrawal H M and Asokan K 2013 *I. J. Electrical and Electronics Engineering* **2** 59
[15] Wagner K W 1973 *Ann. Phys.* **40** 317
[16] Jonker G H 1959 *J. Phys. Chem. Solids* **9** 165

IOP Conf. Series: Journal of Physics: Conf. Series **1086** (2018) 012005 doi:10.1088/1742-6596/1086/1/012005

Effect of low to heavily boron doped pyrolized zinc oxide (ZnO:B) thin films

U R Dash[1], M K R Khan[1], M M Rahman[1] and M Kamruzzaman[2]

[1]Department of Physics, University of Rajshahi, Rajshahi-6205
[2]Department of Physics, Begum Rokeya University, Rangpur- 5400

E-mail: unnatiphru4337@gmail.com

Abstract. Boron doped ZnO thin films have been deposited on glass substrate at temperature (350 ± 5) °C by spray pyrolysis technique and studied structural and physical properties. XRD study reveals polycrystalline hexagonal structure. The crystallite size varies in the range 8 – 44 nm irrespective of crystal planes and doping concentration of B as well. SEM study shows the formation of nano-walls in B doped ZnO thin film. There are some voids in the film, which decreased as the B concentration increased. Temperature dependent resistivity of the films has been measured within the temperature range 300 K to 475 K for different B doped samples, which reveals the semiconducting nature of the films. Activation energies ΔE_1, varies from (0.022 - 0.035) eV in the temperature region (303 K-400 K) and ΔE_2, varies from (0.033-0.188) eV in (400 K - 475 K) depending on Boron concentrations.

1. Introduction

The interest in ZnO is renewed by its increasing demand and high prospects in optoelectronics applications because of its direct band structure, wide band gap (Eg, »3.34 eV at 300 K) and large exciton binding energy of »60 meV [1]. The exciton binding energy is 2.4 times of the thermal energy (k_BT=25 meV) at room temperature (RT).Recently, several groups have carriedout research on partially substituted compounds of ZnO by using various methods for improving the electrical performance. Low-resistivity n-type ZnO could be realized by doping with group IIIA [B, Al, Ga, In] elements. The Al, Ga, and In-doped ZnO thin films have been researched by various groups [2–4]. Although recent work has demonstrated the doping of these group IIIA elements onto ZnO, the formation of boron doped ZnO thin films is still difficult because of complex structure of Boron. Here, we have chosen spray pyrolysis method because pyrolysis is a convenient, low-cost, and rapid method for the deposition of thin films, and has been used for many years for the manufacture of conductive glass. It is also an excellent method for preparing films of semiconductor alloys and complex compounds.

2. Experimental Section

Undopped ZnO and ZnO:B have been deposited on ordinary glass substrate by spray pyrolysis technique at substrate temperature of 350 °C. The most commonly available microscope glass slides were used as substrates in the present work. The solution comes to the spray nozzle through a commercially available medical infusion pipe with controlling system. The deposition solution is then sprayed to the substrate by an air compressor. The working solution was prepared by mixing the appropriate volumes of 0.1M Zinc acetate ($Zn(CH_3.COO)_2\ 2H_2O$) dehydrate and 0.1M Boric acid (H_3BO_3) dissolved in distilled water. Here Zinc acetate and Boric acidwere mixed together in different nominal volume ratios to obtain 0%, 0.2%, 0.6%, 1%, 2%, 3% and 4% Boron doped ZnO thin films. The most commonly used solvent is water. The thickness of the films was measured using Newton's ring method. The structural properties were studied using XRD. The XRD diffraction study of undoped and B doped ZnO thin films on glass substrate has been done with a diffractometer (Model D2 PHASER (BRUKER)), using Cu-k_α radiations of wavelength 1.54184 Å with tube current 10 mA and voltage 30 kV. The peak intensities are recorded corresponding to their 2θ values. The XRD data were taken in the range of 2θ = 20° to 80°, with a step size of 0.02°. The morphological studies were done using SEM. Electrical resistivity of thin films was analyzed using Van-der Pauw's method.

3. Results and Discussion

3.1 X-ray Diffraction Study

The x-ray diffraction (XRD) patterns are shown in figure 1, which clearly show that the films are polycrystalline in nature for all compositions. The characteristic peaks were identified comparing with JCPDS card (Card no. 2100100) and these were at $(2\theta)^o$ = 32.7804, 35.3642, 37.2184, 48.3863, 57.4629 and 68.7872 corresponding to the (*hkl*) planes (100), (002), (101), (102), (110), and (112) respectively. For all composition the possible phase is ZnO. From the XRD patterns it is also seen that the intensity of (100), (002) and (101) planes are nearly equal for undoped sample.

From XRD Pattern it is seen that the intensities of the prominent peaks (100), (002) and (101) are increased with the progression of doping concentration except for 5% doped sample. However, increased of prominent peaks are not systematic. Though there is a systematic increased of intensity of (101) planes is observed with concentration. This means that with the doping of B in ZnO, Crystalline orientation is along [101] direction except for 3% and 5% B doped sample. The intensity of 5% B doped sample is very small which indicates that this sample is poorly crystalline and may be more doping will induce amorphous nature of the sample. The crystalline orientation of 3% and 5% B doped sample is along [002] direction. This type of variation of crystal orientation may be due to the effects of doping concentration and growth temperature as well. Although, in this work a particular temperature 350± 5 °C was used to prepare all the films yet there was a possibility of substrate temperature variation during the growth because temperature control is not an easy task since the heat of substrate comes from cooking heater. So the variation of temperature during growth is inevitable which may change the crystal orientation. Both doping and temperature dependent crystal orientation was reported [5].

The observed peaks positions and d-spacing of the films matched only with Hexagonal ZnO structure imply that there are no secondary phases such as a Boron cluster or oxides. So it is assumed that B^{2+} ions replaced Zn^{2+} sites as substitutional atom in a wurtzite structure.

The XRD pattern suggested that the obtained phases are the characteristic plane of ZnO. The d_{hkl} values were calculated using the Bragg's law,

$$2d\sin\theta = n\lambda \tag{1}$$

where, d is the inter-planer spacing, θ is the Bragg angle and λ is the wavelength of radiation. The corresponding lattice parameters were calculated using the formula:

$$\frac{1}{d^2} = \frac{4}{3}\{(h^2 + hk + l^2)/a^2\} + \frac{l^2}{c^2} \tag{2}$$

The crystallite size was calculated by using Debye-Scherrer formula [2]:

$$\xi = \frac{K\lambda}{\beta\cos\theta} \tag{3}$$

where, ξ is the Crystallite size, K is a constant whose value is 0.94, λ is the wavelength of X-ray used, β is the full width at half maxima (FWHM) expressed in radians and θ is the Bragg angle. The calculated lattice parameter, crystallite size, dislocation density and crystalline strain are given in Table 1.

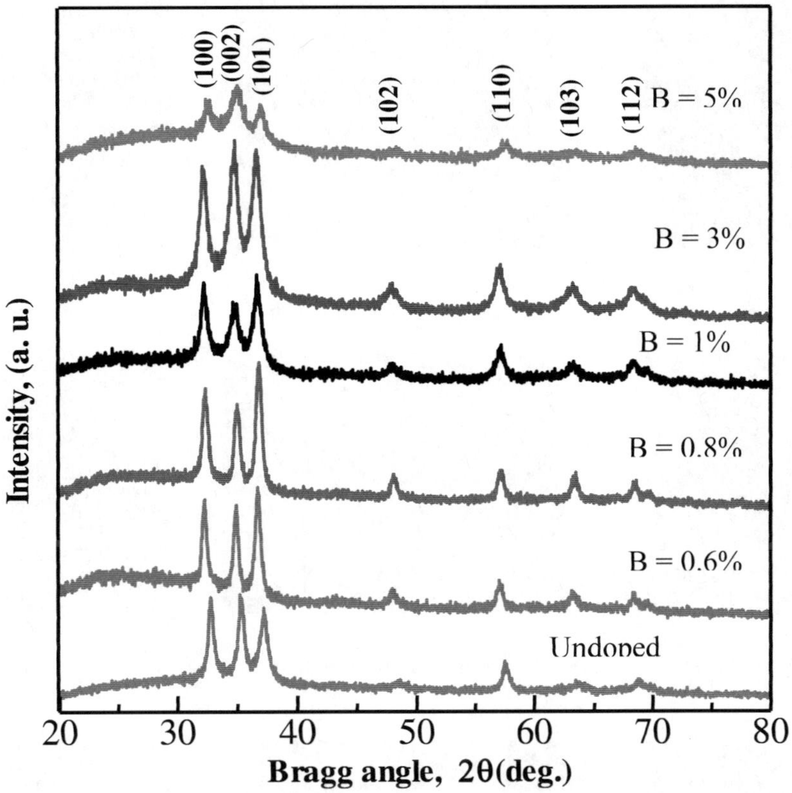

Figure 1. X-ray diffraction patterns of undoped and Boron doped ZnO thin films.

Table-1: Crystal lattice parameter, crystallite size, dislocation density and crystalline strain.

B%	Lattice parameter (nm)		Crystallite size, ξ (nm)		Dislocation density, δ x10^{-3} lines/nm^2		Crystal strain, ϵ x10^{-3}	
	a	c	(100)	(002)	(100)	(002)	(100)	(002)
0.0	0.315	0.507	27.45	18.43	1.33	2.94	4.671	6.464
0.6	0.320	0.514	24.37	24.54	1.63	1.66	5.342	4.918
0.8	0.320	0.511	21.94	27.62	2.07	1.31	5.931	4.343
1.0	0.320	0.514	27.41	15.77	1.33	4.02	7.139	4.759
3.0	0.321	0.515	27.40	13.80	1.33	5.25	4.771	8.772
5.0	3.188	5.130	21.94	10.04	2.08	9.90	5.906	11.989

3.2 Scanning Electron Microscopy study

Scanning electron microscopy (SEM) has been performed by Field emission scanning electron microscope (Philips XL30 FEG) at different magnifications to observe surface morphology and find out the existence of grain, defect, etc. From figure 2(a,b,c) it is seen that poorly dense nanowalls are randomly distributed over the surface. These nanowalls contain small particles with some voids as observed on the nanowalls. As the B concentration increases small particles are uniformly distributed over the surface and the voids are decreased (shown in figure 2(d,e,f)). For 5% B doped sample small particles are distributed in such a way that they make big circles over the surface. Within the circle it is seen that semispherical particles are distributed over the surface (shown in figure 2(g,h,i)).

IOP Conf. Series: Journal of Physics: Conf. Series **1086** (2018) 012005 doi:10.1088/1742-6596/1086/1/012005

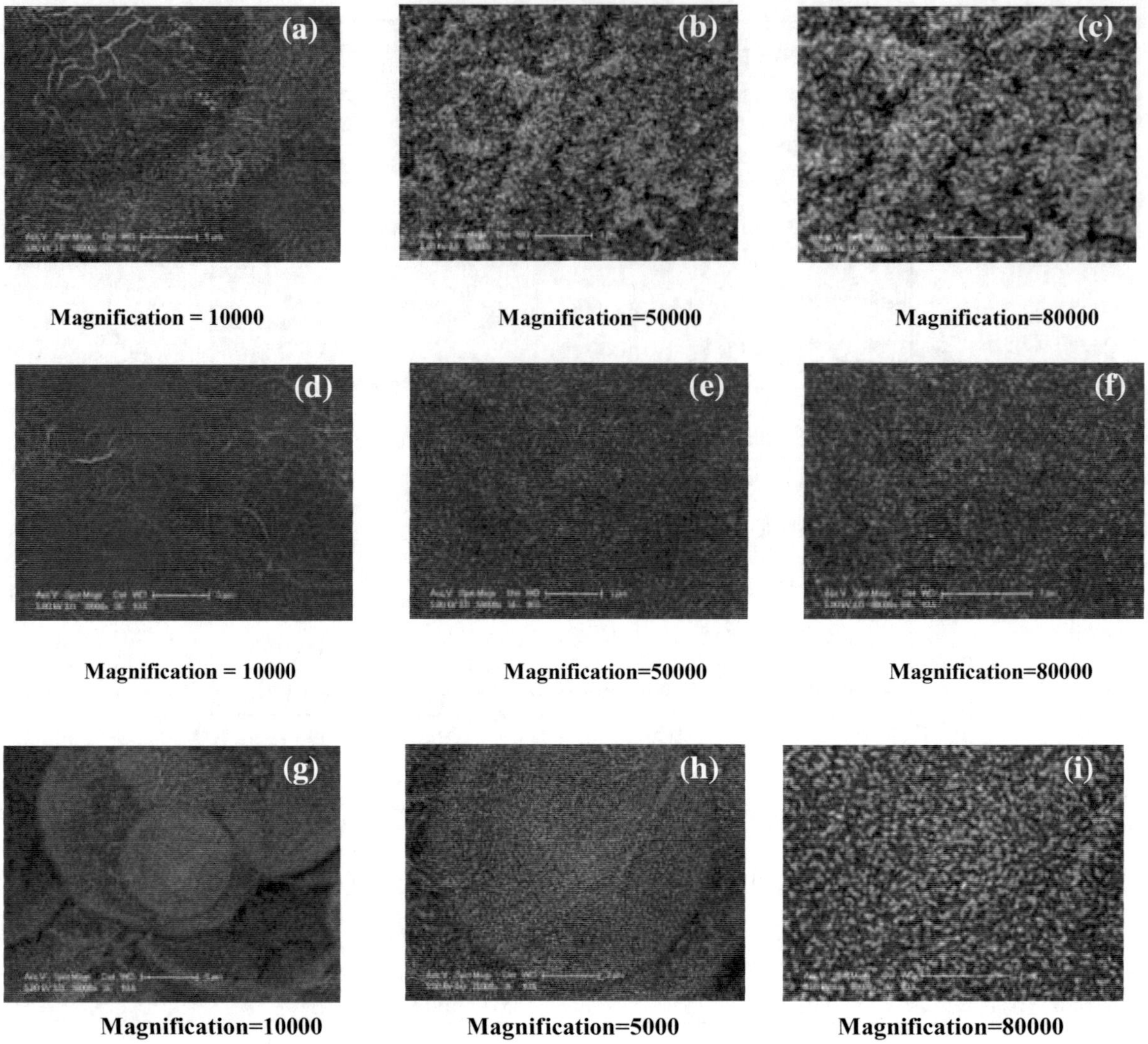

Figure 2. SEM micrographs of 1% (a,b,c), 3% (d,e,f) and 5% (g,h,i) boron doped ZnO.

3.3 Electrical Properties

Variation of resistivity and conductivity with temperature for undoped and Boron doped ZnO thin films are measured by Vander Pauw's [6] method in air ambient in the temperature range (300 K-475 K). From figure 3(a) it is observed that the resistivity decreases with increasing temperature, which is a general feature of a usual semiconductor; both for undoped and Boron doped ZnO thin films. The resistivity was lowest for undoped ZnO thin films which increased with increasing doping concentration of boron. From the figure it is also observed that the room temperature resistivity increases with doping concentration. It is well known that the radius of B is smaller (0.88 Å) compared to Zinc atom (1.33 Å), so during the crystallization it may occupy partially at interstitial position along with substitutional solid solution. Furthermore, it may take the position in grain boundary. Therefore, with boron doping RT resistivity may increase. Conductivity increases with the increase of temperature as shown in figure 3(b). Form this figure it is also seen that conductivity decreases with increasing B doping concentrations.

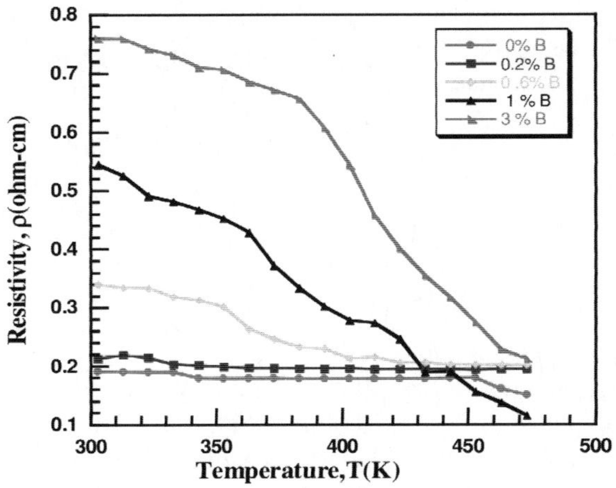

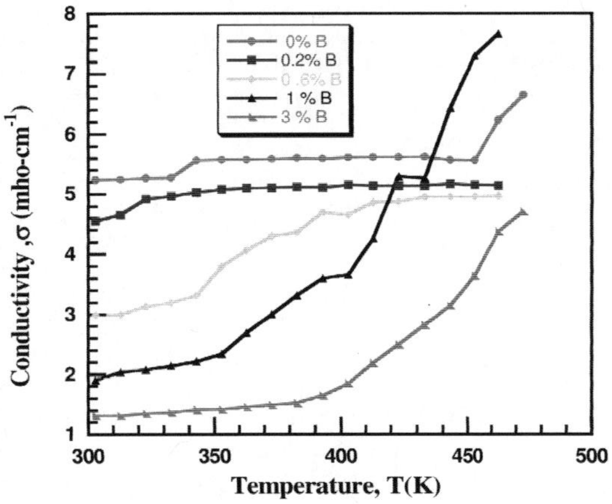

(a) **(b)**

Figure 3. Variation of resistivity and conductivity with temperature for Boron doped ZnO films.

3.4 Activation Energy

The activation energy can be calculated from the equation

$$\sigma = \sigma_0\, e^{-\Delta E/2K_BT} \quad (4)$$

For the calculation of activation energy, ΔE, the lnσ versus 1/T curves for Boron doped ZnO thin films are drawn in figure 4. The activation energies of the films have been calculated from the slopes of the curves. The activation energy of the films is estimated from two regions, ΔE_1 for low (303 K-400 K) temperature region and ΔE_2 for high (400 K-475 K) temperature region. These values are given in Table2. From the table it is observed that the activation energy $\Delta E_2$is higher than that of ΔE_1. The values indicate that conductivity is associated with the impurity band transition of carriers i.e. from donor level to conduction band and the low values of ΔE_1 may be associated with the localized levels hopping due to the excitation of carriers from one defect state to other.

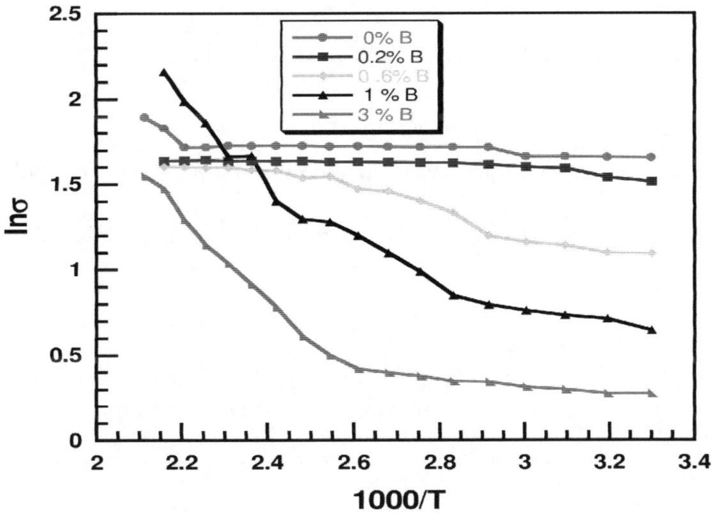

Figure 4. Variation of lnσ with 1/T for Boron doped ZnO thin films.

IOP Conf. Series: Journal of Physics: Conf. Series **1086** (2018) 012005 doi:10.1088/1742-6596/1086/1/012005

Table 2. Activation energies evaluated for undoped and Boron doped ZnO thin films.

Sample	% of Boron	Activation energy, ΔE_1(eV)	Activation energy, ΔE_2(eV)
B doped ZnO	0	0.022	0.033
	0.2	0.032	0.002
	0.6	0.106	0.024
	1	0.112	0.172
	3	0.035	0.188

3.5 Study of Hall Coefficient, Hall Mobility and Carrier Concentration

The Hall voltage measurement was done at room temperature and at constant magnetic field of 9.815KG. From the Hall voltage measurement, Hall constant (R_H), Hall mobility (μ_H) and carrier concentration (n) have been calculated and are tabulated in Table 3. It is seen that Hall coefficient increases with increasing doping concentration of B. The carrier concentration n and conductivity decreases with increasing doping concentration. The calculated values of carrier concentrations belonging to these films are found to be the order of $\sim 10^{16}$. From figure 5 we see that carrier concentration and mobility decreases whereas resistivity increases with increasing B concentration.

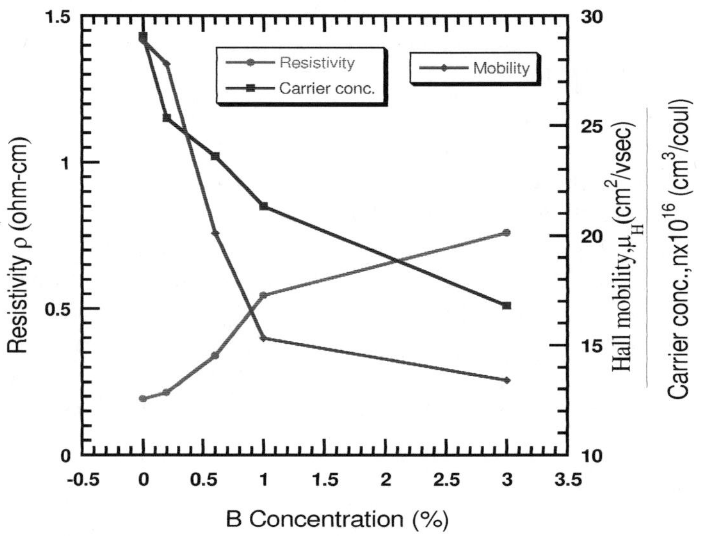

Figure 5. Variation of resistivity, carrier concentration and Hall mobility with Boron concentration.

Table 3. Hall constant (R_H), Hall mobility (μ_H), carrier concentration (n) and conductivity (σ)

4. Conclusion

Doping concentration (B %)	Hall coefficient, R_H (cm^3/coul)	Type	Hall mobility, μ_H (cm^2/Vsec)	Carrier concentration, n x10^{16} (cm^{-3})	Room temperature conductivity, σ (mho-cm^{-1})
0.0	-5.51	n	28.87	1.43	5.24
0.2	-5.93	n	27.81	1.15	4.69
0.6	-6.82	n	20.11	1.02	2.95
1.0	-8.02	n	15.32	0.85	1.91
3.0	-10.16	n	13.41	0.51	1.32

ZnO and B doped ZnO thin films are deposited by spray pyrolysis method. The deposited films are polycrystalline with hexagonal structure as confirmed by XRD experiment. The preferred crystallographic orientation of ZnO in (002) plane changed to (101) by doping concentration of B.

IOP Conf. Series: Journal of Physics: Conf. Series **1086** (2018) 012005 doi:10.1088/1742-6596/1086/1/012005

The SEM micrograph confirms small particles with some voids as observed on the nanowalls. Hall measurements of the B doped ZnO films were carried out at room temperature, which confirmed that the undoped and B doped ZnO thin films are n-type semiconductor. The carrier concentration was found to decrease with increasing doping concentration of B and the carrier concentration was found to the order of $\sim 10^{16}$ cm^{-3}.

Acknowledgement
One of the authors Unnati Rani Dash is thankful to the Ministry of Science and Technology (MOST) of the Peoples Republic of Bangladesh for providing financial support.

References
[1] Liang W Y and Yore A D 1968 *Phys. Rev. Lett.* **20** 59
[2] Wang R, Sleight A W and Cleary D 1996 *Chem. Mater.* **8** 433
[3] Ambrosini A, Malo S and Poeppelmeier K 2002 *Chem. Mater.* **14** 58
[4] Palmer G B, Poeppelmeier K R and Mason T O 1997 *Chem. Mater.* **9** 3121
[5] Liton M N H, Khan M K R, Rahaman M M and Islam M M 2015 *Journal of scientific research* **7(1-2)** 23
[6] Öztas M and Bedir M 2008 *Thin Solid Films* **516** 1703

Boundary layer analysis of momentum heat and mass transfer along a stretching wedge in a nanofluid with magnetic field

M Ali[1] and M A Alim[2]

[1] Department of Mathematics, Chittagong University of Engineering and Technology, Chittagong-4349, Bangladesh.
[2] Department of Mathematics, Bangladesh University of Engineering and Technology, Dhaka-1000, Bangladesh.

E-mail: ali.mehidi93@gmail.com, E-mail: maalim@math.buet.ac.bd

Abstract. The effect of various dimensionless parameters on boundary layer flow are analysed. The MHD nanofluid flow along a moving stretching wedge surface are considered. The governing partial differential equations are transformed into ordinary differential equations by using the similarity transformation. These ordinary differential equations are numerically solved using fourth order Runge–Kutta method along with shooting technique. The results indicate that the velocity profiles increases with increasing values of pressure gradient parameter β for different situation and also for magnetic parameter M and for stretching parameter λ. The temperature profiles decreases for velocity ratio parameter λ, Brownian motion parameter Nb and Prandtl number Pr but opposite result arises for increasing values of thermoporesis parameter Nt. The concentration decreases with an increase in pressure gradient parameter β, velocity ratio parameter λ, Brownian motion parameter and Lewis number Le, but increases for thermophoresis parameter. Finally, for validity and accuracy the present results are compared with previously published work.

1. Introduction

The main application of the boundary layer theory is the calculation of the skin friction drag when the body is moving or static in a fluid, such as the drag of a flat plate, an airplane wing, a turbine blade, or a complete ship. In case of convective heating or cooling purpose, such as, cooling of an engine and cooling of a hot plate in case of an external flow condition, a fluid boundary layer is formed that gives rise to thermal boundary layer. Besides moving or stretching surface is important in many engineering processes such as production of polymeric sheets, paper production, wire drawing, insulating materials, drawing of plastic films and fine-fiber mats etc. So on the basis of the Prandtl boundary layer theory, Falkner – Skan developed a model which is not parallel to the flow known as wedge flow. Yacob et al. [1] discussed the Falker- Skan flow along a static / moving wedge in a nanofluids, Hayat et al. [2] discussed the Falkner- Skan flow in case of power – law fluid with mixed convection, and Ashwini et al. [3] studied the unsteady MHD accelerating flow past a wedge with thermal radiation and internal heat generation. Again the concept of nanofluid was first introduced by Choi in 1995 which referring to dispersions of nanoparticles in the base fluids such as water, ethylene glycol, and propylene glycol. Later, Buongiorno [4] first examined the reasons behind the enhancement in heat transfer rate in case of nanofluid flow. So, he observes that the Brownian diffusion and thermophoresis are the main causes to enhance the heat transfer rate, Nield and Kuznetsov [5] and Kuznetsov and Nield [6] investigated the natural convective boundary layer flow of a nanofluid using

Buongiorno model. Besides many researchers [7-10] worked related to this boundary layer theory by considering various types of geometry and boundary conditions. By applying the model of Nield and Kuznetsov [5], Khan and Pop [11] first studied the boundary layer nanofluid flow past a linearly stretching sheet. Mutuku [12] discussed MHD boundary layer flow over a permeable vertical plate with convective heating. Later, Makinde et al. [13] discussed the combined effects of buoyancy force and magnetic field on stagnation-point flow and heat transfer in a nanofluid flow towards a stretching sheet. Hence, the present work has been focused on the steady MHD Falkner-Skan nanofluid flow with magnetic field past a stretching wedge. Thus, in section 2, the mathematical models of boundary layer flow have been presented. The results and discussion has been described in Section 3. Finally, the conclusion has been drawn in section 4.

2. Governing equations and similarity analysis

Let us consider the wedge is moving with a velocity $u_w(x,t)$ and the free stream velocity is $U(x,t)$. Also consider the flow is two dimensional steady laminar and viscous incompressible electrically conducting. The x-axis is measured along the surface of the wedge and the positive y-coordinate is measured normal to the x-axis in the outward direction towards the fluid. The temperature of the wedge wall T_w and nanoparticle concentration C_w are variable and the free stream temperature and nanoparticle concentration are T_∞ and C_∞ respectively far away from the boundary layer, the total angle of the wedge is denoted as $\beta\pi$, where β is the Hartree pressure gradient. It is assumed that the base fluid and the nanoparticles are in thermal equilibrium and no slip occurs between them. Also, a uniform magnetic field of strength B0 is introduced to the normal to the direction of the flow.

Under these assumption, along with boundary layer approximation and neglecting the viscous dissipation, the governing partial differential equations for the flow field are given by [4]:

Equation of continuity:

$$\frac{\partial u}{\partial x}+\frac{\partial v}{\partial y}=0 \tag{1}$$

Momentum equation:

$$u\frac{\partial u}{\partial x}+v\frac{\partial u}{\partial y}=U\frac{dU}{dx}+\nu_f\frac{\partial^2 u}{\partial y^2}+\frac{\sigma B_0^{\,2}}{\rho_f}(U-u) \tag{2}$$

Energy equation:

$$u\frac{\partial T}{\partial x}+v\frac{\partial T}{\partial y}=\alpha_f\frac{\partial^2 T}{\partial y^2}+\tau\left\{D_B\left(\frac{\partial T}{\partial y}\frac{\partial C}{\partial y}\right)+\frac{D_T}{T_\infty}\left(\frac{\partial T}{\partial y}\right)^2\right\} \tag{3}$$

Nanoparticle concentration equation:

$$u\frac{\partial C}{\partial x}+v\frac{\partial C}{\partial y}=D_B\frac{\partial^2 C}{\partial y^2}+\frac{D_T}{T_\infty}\frac{\partial^2 T}{\partial y^2} \tag{4}$$

IOP Conf. Series: Journal of Physics: Conf. Series **1086** (2018) 012006 doi:10.1088/1742-6596/1086/1/012006

The above equations are subject to the following boundary conditions:

$$u = u_w = ax^m,\ v = 0,\ T = T_w - T_\infty = cx^s, C = C_w - C_\infty = dx^s\ at\ \ y = 0$$
$$u = U = bx^m, T \to T_\infty, C \to C_\infty\ as\ \ y \to \infty$$

where u and v are the velocity components along x and y directions, T_w and C_w are the variable wall temperature and nanoparticle volume fraction of the wall while the uniform temperature and nanoparticle volume fraction far from the wall are T_∞ and C_∞ respectively, ν_f is the kinematic viscosity of the base fluid, ρ_f is the density of the base fluid, σ is the electrical conductivity, B_0 is the magnetic field intensity, g is the acceleration due to gravity, α_f is the thermal diffusivity of the base fluid, D_B is the Brownian diffusion coefficient, D_T is the thermophoresis diffusion coefficient. Here τ is the ratio of the effective heat capacity of the nanoparticle material and the heat capacity of the ordinary fluid, T is the fluid temperature and C is the nanoparticle volume fraction respectively.

Where a, b, c and d are positive constant with dimension reciprocal time. Among them, a and b are the initial stretching rate, and the exponent m is a function of the wedge angle parameter β where the total apex angle of the wedge is βπ such that $\beta = \frac{2m}{1+m}$. To convert the governing equations into a set of ordinary differential equations, the following similarity transformations are applied:

$$\eta = y\sqrt{\frac{U(1+m)}{2x\nu}},\ \psi = \sqrt{\frac{2x\nu U}{(1+m)}}f(\eta), \theta(\eta) = \frac{T - T_\infty}{T_w - T_\infty}, \varphi(\eta) = \frac{C - C_\infty}{C_w - C_\infty}, u = \frac{\partial\psi}{\partial y}\ \ and\ v = -\frac{\partial\psi}{\partial x}$$

By applying the above similarity transformations, the partial differential equations (2), (3) and (4) are transformed into non-dimensional, nonlinear and coupled ordinary differential equations:

$$f''' + ff'' + \beta\left(1 - f'^2\right) + \frac{2M}{1+m}(1 - f') = 0 \tag{5}$$

$$\theta'' + \Pr\left[f\theta' - \frac{2s}{1+m} f'\theta + Nb\theta'\varphi' + Nt\theta'^2 \right] = 0 \tag{6}$$

$$\varphi'' + \frac{Nt}{Nb}\theta'' + Le\left[f\varphi' - \frac{2s}{1+m} f'\varphi \right] = 0 \tag{7}$$

The transformed boundary conditions:

$$f = 0\ f' = \lambda, \theta = 1, \phi = 1\ at\ \eta = 0\ \text{and}\ f' \to 1, \theta = \phi \to 0\ \ as\ \eta \to \infty$$

$$M = \frac{\sigma B_0^2 x}{\rho_f U}, \lambda = \frac{a}{b}, \Pr = \frac{\nu_f}{\alpha_f}, Nb = \frac{\tau D_B (C_w - C_\infty)}{\nu_f},$$

$$Nt = \frac{\tau D_T (T_w - T_\infty)}{T_\infty \nu_f}, \beta = \frac{2m}{1+m}, Le = \frac{\nu}{D_B}$$

Where M is the magnetic parameter, λ is stretching parameter, Pr is the Prandtl number, Nb is the Brownian motion, Nt is the thermophoresis parameter, β is the pressure gradient and Le is Lewis number respectively. The important physical quantities of this problem are skin friction coefficient C_f , the local Nusselt number *Nu* and the local Sherwood number *Sh* which are proportional to rate of velocity, rate of temperature and rate of nanoparticle concentration respectively.

3. Results and Discussion

The nonlinear coupled ordinary differential equations along with boundary conditions are solved numerically using Runge –Kutta 4th order integration scheme with shooting method. The effect of various parameters on velocity, temperature and concentration are shown in graphically and also in tabular form. The present results are compared with previously published work which are shown in Table1. This ensures the validity and accuracy of the present work. Also the various values of skin friction, rate of heat transfer and rate of mass transfer are presented in Table 2 for different values of *Nb, Nt,* λ and Le by taking $Pr = 1.0$ and $M = 0.1$.

The effect of pressure gradient parameter on velocity profile are shown in Fig.1 (a) – Fig.1 (d). From Fig.1 (a) and Fig.1 (b) the velocity profiles increases with increasing values of pressure gradient parameter β in absence of magnetic parameter and stretching parameter and also in presence of stretching parameter. Therefore in absence of magnetic effect and taking static wedge it is seen that the separation is occurred at β = -0.198. But in case of moving wedge and without the effect of magnetic field, the separation is occurred at β = -0.35. From Fig.1 (a) and Fig.1 (d) it is observed that the velocity profiles increases for increasing values of magnetic parameter and stretching ratio parameter because the free stream velocity is dominating here.

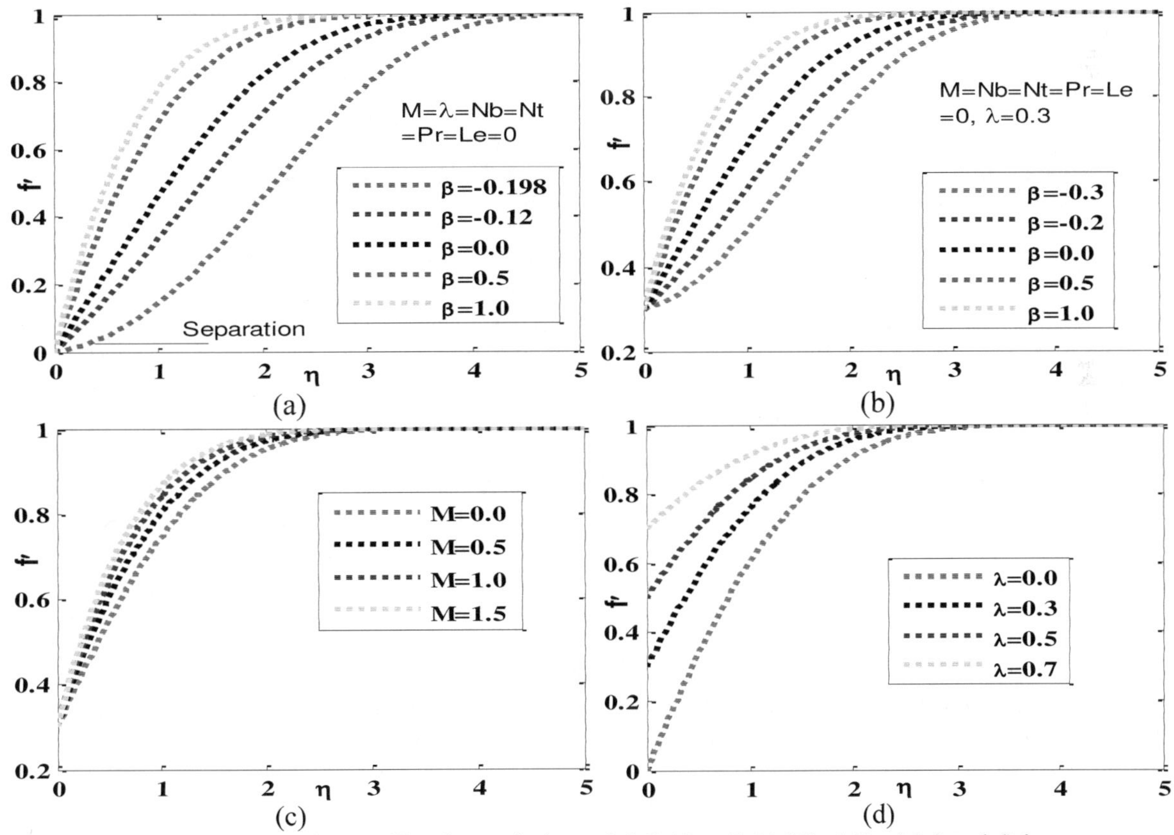

Figure 1. Velocity profiles for variation of a) β,(λ = 0) b) β(λ ≠ 0), c) M and d) λ

From Fig.2 (a) – Fig.2 (d), the temperature profiles increases for thermophoresis parameter but reverse results arises for other entering parameters. The increment of Prandtl number results in major effects

IOP Conf. Series: Journal of Physics: Conf. Series **1086** (2018) 012006 doi:10.1088/1742-6596/1086/1/012006

on temperature profiles. The thermal boundary layer thickness reduces with Prandtl number and it happens due to decrease of thermal diffusivity for the increment of Prandtl number.

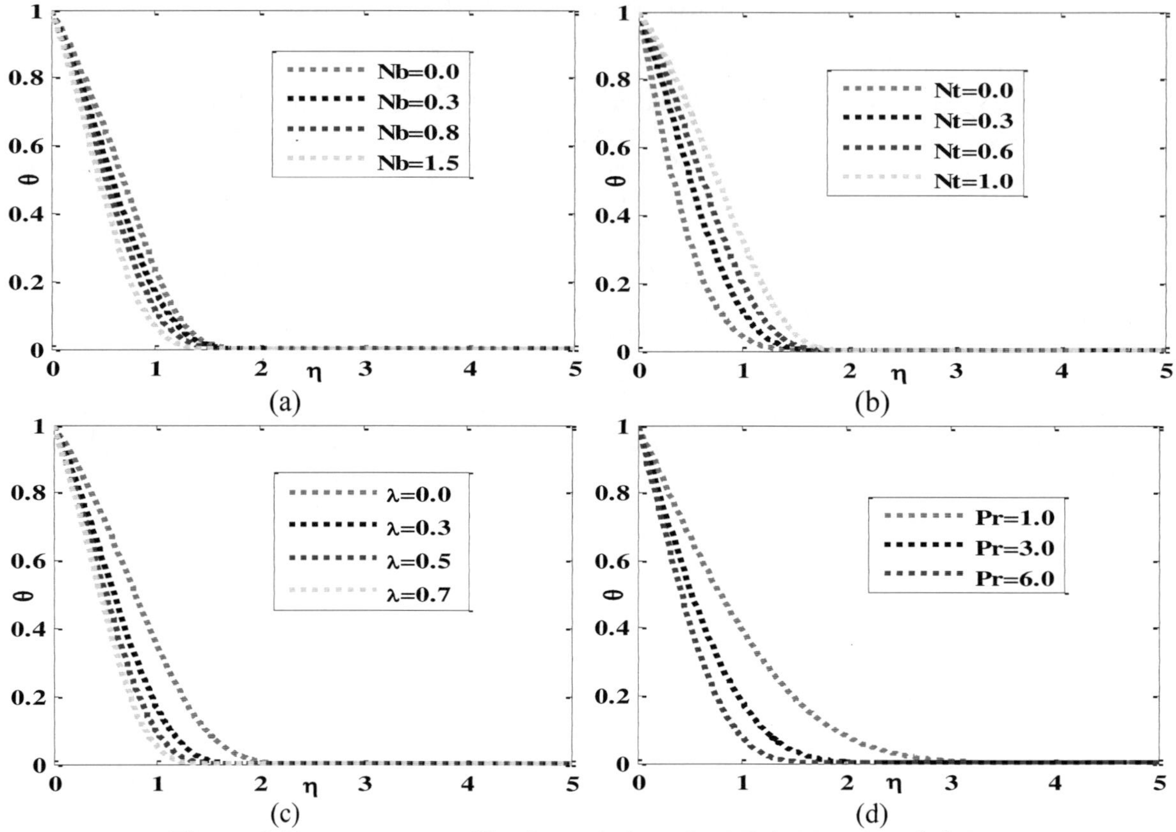

Figure 2. Temperature profiles for variation of a) Nb b) Nt c) λ and d) Pr

From Fig. 3(a) – Fig.3 (d) it is observed that the concentration profiles increases for thermophoresis parameter but reverse results arises for the remaining parameters. Thermophoresis parameter is a key parameter for analysing the temperature distributions and nanoparticles concentration in nanofluid flow. The effect of thermophoresis parameter on the concentration profiles is presented in Fig.3 (b). Therefore, for increasing values of thermophoresis parameter, the temperature profiles and the concentration profiles in the boundary layer increases. Increase in thermophoresis parameter causes increment in the thermophoresis force which tends to move nanoparticles from hot to cold areas and consequently it increases the magnitude of temperature profiles and nanoparticle concentration profiles. Ultimately, the thickness of concentration boundary layer becomes significantly large for slightly increasing values of thermophoresis parameter. The large values of Le, the concentration significantly decreases and as a result the thickness of concentration boundary layer reduces. But, for smaller values of Le the overshoot is found near the wedge. The Brownian diffusion effect becomes nominal for larger values of Lewis number and for which the concentration boundary layer thickness decreases.

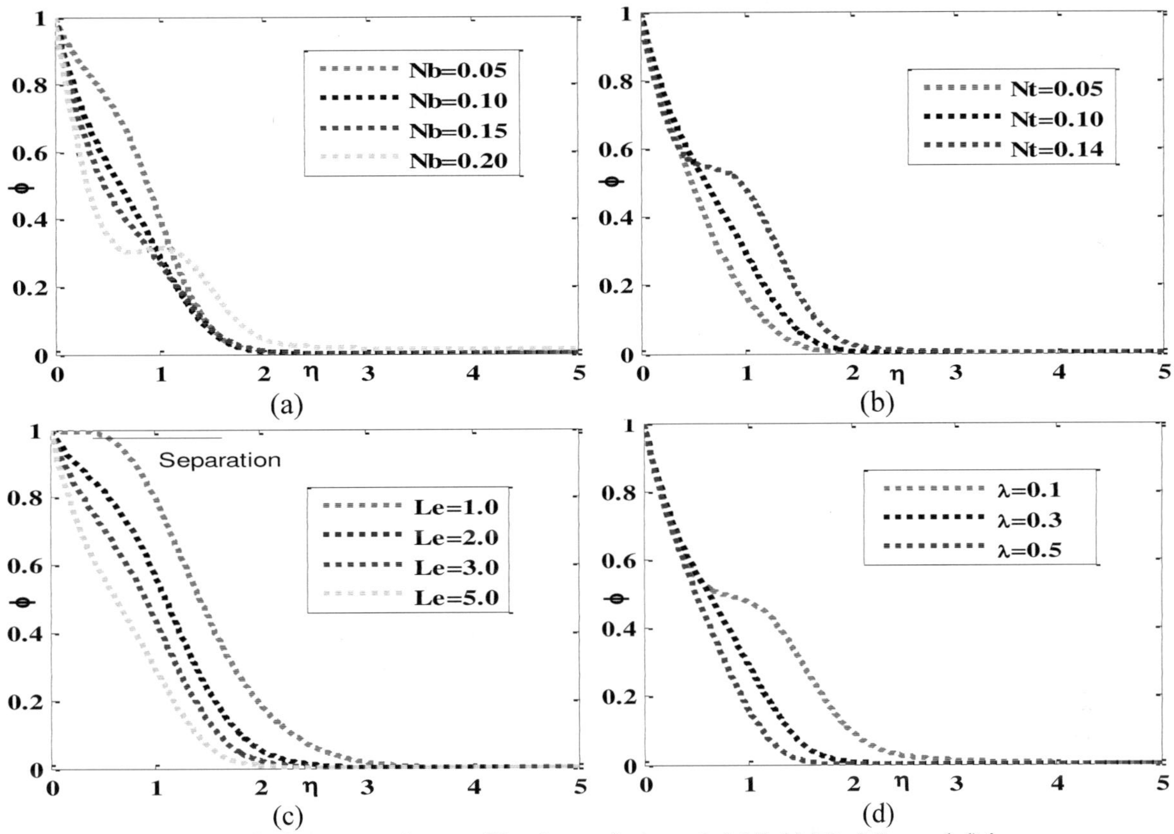

Figure 3. Concentration profiles for variation of a) Nb b) Nt c) Le and d) λ

Table 1. Comparison of skin friction [$f''(0)$] for different values of β, when M = Pr = Nb = Nt = Le = λ = 0.

	Rajagopal et al. [14]	White [15]	Mohammadi et al. [16]	Khan & Pop [17]	Present results
β	$f''(0)$	$f''(0)$	$f''(0)$	$f''(0)$	$f''(0)$
-0.12	-	-	0.281772	-	0.28211
-0.15	-	-	0.216335	-	0.217153
-0.18	-	-	0.128637	-	0.13138
0.0		0.4696	0.469589	0.4696	0.46964
0.2	0.686708	-	-	-	0.686690
1/6	-	0.6550	-	0.6550	0.6550
1/3	-	0.8021	-	0.8021	0.80212
0.5	0.927680	0.9277	0.927601	0.9277	0.92768
2/3	-	1.0389	-	1.0389	1.0389
1.0	1.232585	1.2326	1.232587	1.2326	1.232587
1.6	1.521514	-	-	-	1.52151399

Table 2. Values of skin friction [$f''(0)$], local Nusselt number $-\theta'(0)$ and local Sherwood number $-\varphi'(0)$ for different values of Nb, Nt, λ when Pr = 1.0 and M = 0.1.

λ	Nb	Nt	Le	$f''(0)$	$-\theta'(0)$	$-\varphi'(0)$
0.3	0.4	0.4	2.0	0.59710	0.64858	0.70207
0.5	0.4	0.4	2.0	0.45634	0.76658	0.78802
0.7	0.4	0.4	2.0	0.29051	0.87148	0.92169
0.3	0.8	0.4	2.0	-	0.71569	0.75069
0.3	1.5	0.4	2.0	-	0.80455	0.78123
0.3	0.4	0.3	2.0	-	0.83799	0.70722
0.3	0.4	0.6	2.0	-	0.58416	0.74463
0.3	0.4	0.4	1.0	--	-	0.45265
03	0.4	0.4	3.0	-	-	0.87037

4. Conclusion

According to the Falkner-Skan model the solution is exist for $-0.198 \leq \beta \leq 2$. But the present solution is exist for $-0.35 < \beta \leq 2$ which are shown in Fig. 1(b). The velocity increases for magnetic parameter because the free stream velocity U is dominating here and the velocity decreases for stretching ratio parameter but the thickness of momentum boundary layer increases. Again, the separation is observed for Nb < 0.2, Nt < 0.14, Le > 1.0 and λ > 0.1 in the concentration profiles.

References

[1] Yacob A N, Ishak A and Pop I 2011 *I. J. Ther. Sci.* **50** 133
[2] Hayat T, Majid H, Nadeem and Meslou S 2011 *Comput. Fluid* **49** 22
[3] Ashwini G and Eswara A T 2015 *I. J. Math. Comput. Sci.* **1** 13
[4] Buongiorno J 2006 *J. Heat Transfer* **128** 240
[5] Nield D A and Kuznetsov A V 2009 *I. J Heat Mass Transfer* **52** 5792
[6] Kuznetsov A V and Nield D A 2010 *I. J. Ther. Sci.* **49** 243
[7] Ali M, Alam M A and Alim M A 2016 *Bangladesh J. Sci. Industrial Res.* **51** 139
[8] Ali M, Alam M A, Chowdhury M Z U and Alim M A 2016 *J. Sci. Res.* **8** 119
[9] Ali M, Alam M A, Chowdhury M Z U and Alim M A 2016 *Int. J. Mech. Eng. Autom.* **3** 144
[10] Alm M A, Ali M and Alim M A 2016 *J. Sci. Res.* **8** 41
[11] Khan W A and Pop I 2010 *I. J. Heat Mass Transfer,* **53** 2477
[12] Mutuku-Njane W N and Makinde O D 2014 *J. Comput Theoretical Nano Sci.* **11** 667
[13] Makinde O D, Khan W A and Khan Z A 2013 *I. J. Heat Mass Transfer* **62** 526
[14] Rajagopal K R and Gupta A S 1983 *I. J. Non Linear Mech.* **18** 313
[15] White F M 1991 2nd edn. McGraw-Hill, New York, NY, USA
[16] Mohammadi F, Hosseini M M, Dehgahn A and Maalek Ghaini F M 2012 *Studies in Nonlinear Sci.* **3** 86
[17] Khan W A and Pop I 2013 *Math. Prob. Eng.* **1**

IOP Conf. Series: Journal of Physics: Conf. Series **1086** (2018) 012007 doi:10.1088/1742-6596/1086/1/012007

Perception of thermal comfort and prediction of heat stress (PHS) due to heat gain in tropical university classrooms

B K Biswas[1, 2] and Y Hamada[2]

[1]Department of Architecture, Chittagong University of Engineering and Technology, Bangladesh
[2]Research Group of Planning and Performances for Built Environment, Division of Human Environmental Systems, Faculty of Engineering,Hokkaido University, Kita-u, Sapporo, Hokkaido

E-mail: kantibiplob@gmail.com, hamada@eng.hokudai.jp

Abstract. Comfortable indoor climate is anticipated in hot-humid tropical condition. University classrooms are such kind of space which quality can improve the occupant's performance directly. Due to lack of their adaptation ability, like the fixed schedule of activities and clothing causes extra heat gain from the environment and make hazardous situation. In this research, we focused to find out the comfortable condition of university classrooms in Bangladesh. The field monitoring was conducted at Chittagong University of Engineering and Technology (CUET) during the hottest months in the country (April-July). The measurements included the basic thermal environmental parameters like air temperature (AT), relative humidity (RH), indoor air velocity (V), and indoor Light (L) during regular class time 9:00to 17:00. Predicted Heat Strain (PHS) has been calculated using obtained data with a known metabolic rate and clothing. It is found that the skin temperature crosses 34°C frequently in PHS scale and creates health risk.

1. Introduction

According to university grants commission of Bangladesh (UGC) [1], governing body of universities in Bangladesh, there are 37 government universities in the country. University student's performance including attention, concentration, learning and hearing significantly depends on a healthy and comfortable indoor climate [2-4]. But most of the cases the indoor climate is not well maintained or designed for the students. Their performance is overlooked according to health impact analysis [5] like other tropical countries. For a vast population as well as financial barriers, it is hard to maintain a proper healthy indoor condition. But its impact is phenomenal. Students have a small adaptive opportunities comparing to residential occupants [6-7].They have scopes to add or remove clothes, control windows and can adjust sun shading devices [8-11].So heat stress occurred and it paramount discomfort and reduce the performance of the students [12-14].From field monitoring, it was observed that in hot humid tropical climate, university classrooms are highly affected by the ambient environment. Basic thermal parameters like air temperature (AT) °C, relative humidity (RH) %, air velocity (v) ms^{-1} and indoor light (lux) are not properly maintained as per thermal comfort defined by ASHRAE [15].Students need to spend a lot of time in classrooms due to their fixed schedules, as a result, their heat gain is constant without any significant adaptation and it indicates a health issue.

Our whole prediction process was conducted by a series of field monitoring on a regular basis. The purpose of this study was to figure out the thermal condition in classrooms by objective measurement and comparison with ASHRAE standards [15] for the hottest months of the year April to July. The objectives are:

- Figure out the heat gain and heat loss factors in indoor climate for heat balance.
- Find out the basic thermal condition in university classrooms during class time at summer season.

- Calculating predicted heat strain (PHS) [16], skin temperature (t_{sk}) to determine the risk in the classrooms and evaluating the safety of students.

2. Methodology

The aim of this study is to investigate the present indoor climatic condition of the university classrooms by means of the objective as well as subjective approach and relate it to the comfort parameters. For figuring out the thermal comfort, it is important to create a relationship between the indoor environments with the comfort parameters. The field study is conducted in a university classroom at the Chittagong University of Engineering and Technology (CUET), Bangladesh. Because of its close proximity to the Bay of Bengal, its location can be identified as the hottest and humid zone in the country. So it shows a vulnerable climatic condition due to comfort analysis. Figure 1 shows the different steps of the research.

2.1. Classroom selection

For selecting classroom it is an important consideration that the building should be constructed of brick and concrete, which is common material all over the country. Another consideration is the climatic complexity. In Chittagong, the climate is mostly hot and humid because of the sea Bay of Bengal is very near to it on the west side. For that reason, the humidity goes higher from another part of the country. Our field monitoring was conducted on the 3rd floor (East part) of a 12 storied building. The building is surrounded by small hills at the east side, at west a 3 storied brick building, and north as well as south parts are open with green trees. Table 1 shows the basic information about study zone. Figure 2 shows the basic layout of the studied space (plan) and Figure 3 shows the longitudinal section of it. The building has a natural ventilation (NV) system with windows and mechanical ventilation (MV) system is controlled by electric ceiling fans. In Bangladesh, most of the classrooms have the same kind of mixed ventilation system.

2.2. Data collection

Our study was conducted during the regular class time started from 9:00 to 17:00 with a 1-hour interval in 2016 during the hottest month of the country April, May, June and July. Measurement was taken at the point shown in Figure 2 and Figure 3. Picture of the classroom and sampling point are shown in Figure 4(a). Indoor air temperature, relative humidity, and air velocity were measured by LM-8102 and Digital Thermo-Hygrometer (Figure 4(b)).Table 2 shows the detail information on these two instruments. After taking data from both of the tools, they were averaged for minimizing the operating errors. At the sampling point, the tools were left to run 3 minutes before taking the values and parameters were recorded. Clear height of the classroom was 2.9meters. All data were recorded at1.1 meters height of the floor, which recommended the seated level of the occupants [23]. Here metabolic rate and clothing insulation were considered from ASHRAE standards [24]. For this study metabolic rate was taken 1.1 met, where students were seated and involved with some light activities such as reading, writing, speaking etc. Clothing of the subject was considered by checklist. For male, it was long pants and shirt or T-shirt with underwear and in case of female, mostly salwar kameez and underwear. The average clothing insulation was found 0.53clo. For determining the clothing insulation, subjects were asked to fill about their wearing. After summing and averaging the value was 0.53 clo according to ASHRAE standards [15].

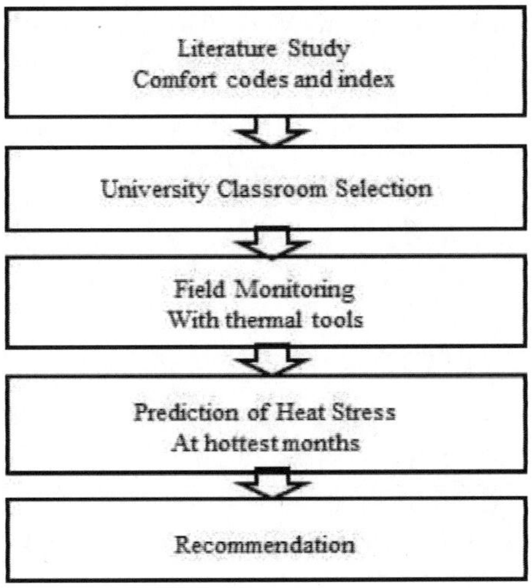

Figure 1. Different steps of the present research

Table 1. Characteristics of the study space

Input consideration	
Room area	271.45 m^2
Room length	30.5 m
Room width	8.9 m
Room height	2.9 m
Building material	Brick, Concrete, Plaster
Occupants	30 (Male 15, Female 15)
Window size	1.22 m X 1.22 m
Window ratio against wall	1 : 9.6
No. of Fan	15 nos. (75 W, air delivery 250 m^3/min)
No. of Light	40 nos. (60 W)

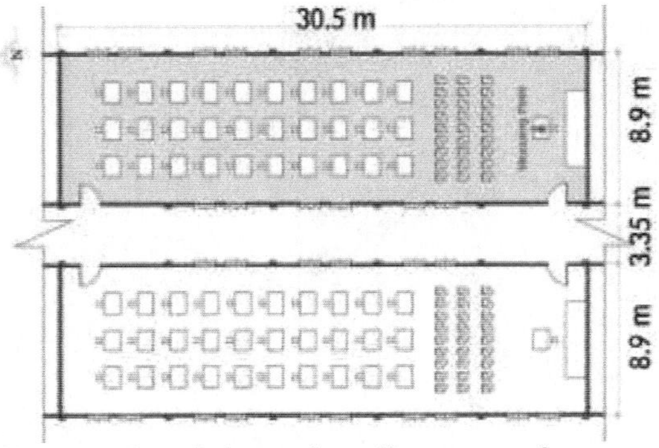

Figure 2. Schematic diagram of space allocation of present study (university classroom plan)

Figure 3. Space view for present study (longitudinal section)

Figure 4(a). Measurement of thermal properties in classroom

Figure 4(b). Experimental tools

For this research, a group of 30 students was performed in the classroom. 15 were male and 15 were female out of them. Average ages of the subjects were between 21 to 23 and had a good health to commit the study works. Table 3 shows the information of the subjects. Subjects were uniformly distributed during the class hours in their classroom. The recorded information was elaborated to evaluate heat stress.

3. Results analysis and discussion

From field monitoring air temperature (AT) °C, relative humidity (RH) %, air velocity (v) ms^{-1} and indoor light (lux) were recorded. Using these data skin temperature (t_{sk}) and core body temperature (t_{cr}) are calculated.

3.1. Basic thermal condition

For human body thermal balance is important to maintain comfort level and reduce heat stress. Core body temperature must remain balanced and constant around 37°C and skin temperature around 34°C. All surplus heat must be released to the environment. Body heat gain occurred by the metabolism (Met) of basal and muscular tissue, conduction (Cnd) with warm subjects, convection (Cnv) process if the air is warmer than the skin, and by

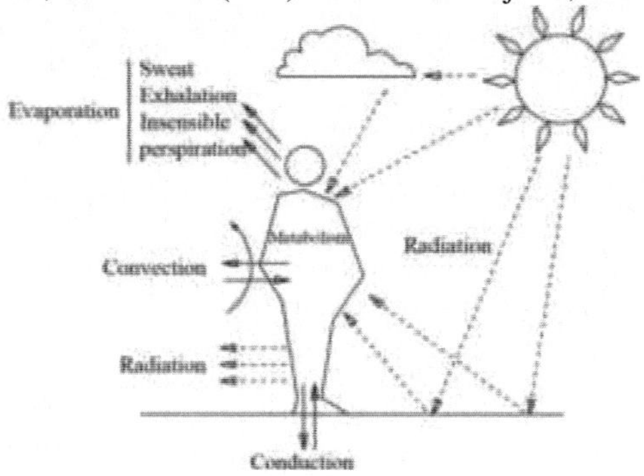

Figure 5. Heat gain and heat loss process

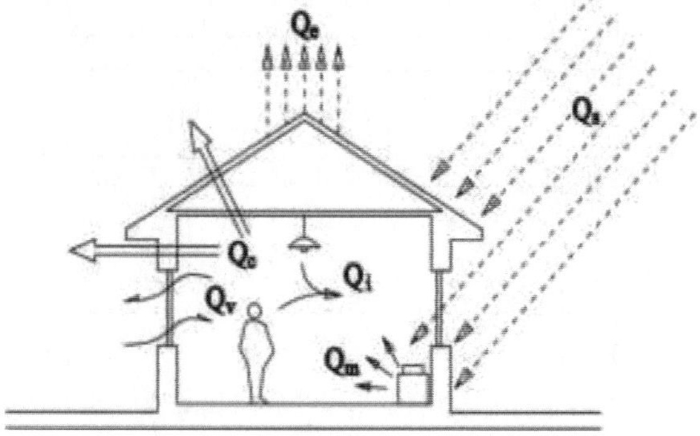

Figure 6. Heat exchanges of buildings

Table 2. Experimental tools

Name of the Tool	Specification
LM-8102(Lutron Electronic Enterprise)	Model: LM-8102 * Sound level meter: 30 to 130 dB * Humidity: 10 to 95 %RH * Anemometer: 0.4 to 30 m/s * Light: 0 to 20,000 Lux *Thermometer: -100-1300 ^{o}C * Sound level meter meet IEC 61672 TYPE 2
Digital Thermo Hygrometer	* Temperature measurement. * Humidity measurement.

Table 3. Description of subjects

Input consideration	
Total occupants	30 nos.
Male	15 nos.
Female	15 nos.
Age	21-23 years
Health condition	Good
Male cloths	Pants and shirt or T-shirt, underwear.(0.53 clo)
Female cloths	Salwar kameez, underwear.(0.53clo)
Activities	Listening lectures, reading, writing
Metabolic rate	1.1 met

radiation (Rad) from the sun, sky or hot bodies. Where heat loss process is conduction (Cnd) with cold subjects, convection (Cnv) process if the air is cooler than skin, radiation (Rad) to night sky or cold surfaces and by evaporation (Evp) process like sweating.

So when the sum of the heat gain and heat loss is "0" it indicates thermal balance. When the heat gain is higher than the heat loss it increases core body temperature and skin temperature and causes discomfort.

$$\text{Thermal balance} = \text{Met} - \text{Evp} \pm \text{Cnd} \pm \text{Cnv} \pm \text{Rad} = 0 \quad (1)$$

Figure 5 shows the heat gain and heat loss process to the human body from the ambient environment. Figure 6 shows the heat exchange process of buildings. Conduction of heat may occur either inward or outward side depending on temperature and here denoted as Q_c. Solar radiation passes through the clear or opaque window and denoted as Q_s. Movement of air creates convection heat flow and denoted as Q_v. Where Q_m is the heat production of outside energy supply in the space. Q_i indicates the internal heat gain from human bodies, lamps, motors and appliances. And if evaporation occurs on the surface of the building and the vapours are removed, this produces cooling effect and denoted as Q_e.

$$\text{So, thermal balance} = Q_i + Q_s \pm Q_c \pm Q_v \pm Q_m - Q_e = 0 \quad (2)$$

In Figure 7 air temperature is plotted from the field study. Every line represents a full day temperature character. Measurement was conducted during regular class time 9:00 to 17:00 in 2016 (April- July). Data were collected in every one hour interval. It shows that ambient air temperature

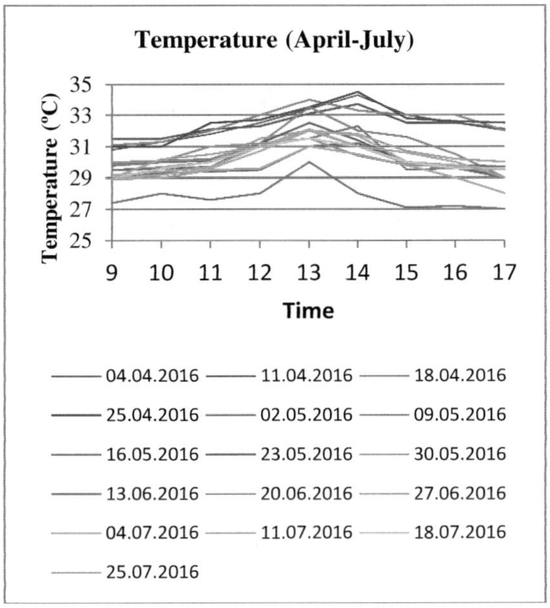

Figure 7. Observed temperature (April- July)

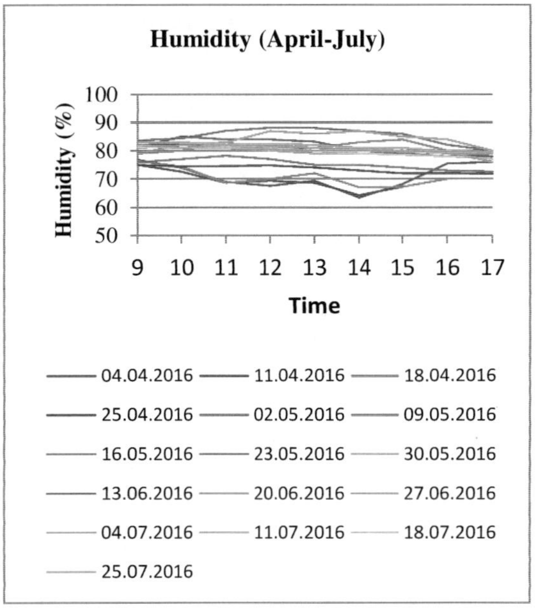

Figure 8. Observed Humidity (April- July)

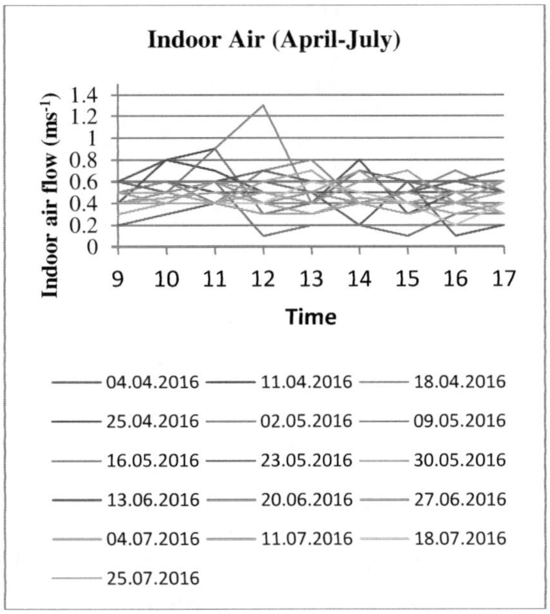

Figure 9. Observed indoor air flow (April-July)

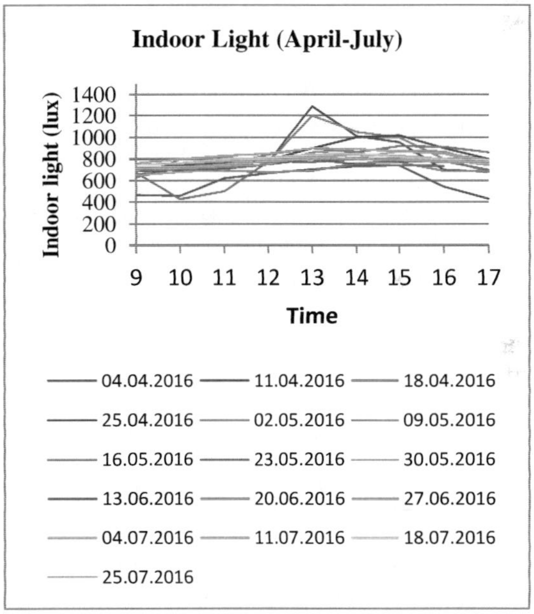

Figure 10. Observed indoor light (April- July)

range travels 27 °C to 35 °C. Where during 12:00 to 15:00 the range is higher and most of the cases it is above skin temperature and heat gain occur due to convection process.

Figure 8 shows the humidity of the air. Data collecting process was same. High humidity restricts heat loss and creates discomfort and it shows that in the study space it is almost 75-85% humidity every time. It increases the skin and core body temperature.

Figure 9 shows the air flow in the study space. Air ventilation helps to exchange the hot air from the

Table 4. Subjective consideration for PHS calculation

Input consideration	unit	value
Height (h)	m	1.72
Weight (m)	kg	61
Body mass index (BMI)	m^2	1.50
Metabolic rate (M)	met	1.1
Clothing rate (I_{clo})	clo	0.53
Rectal Temperature (t_{re})	^oC	38.00
Moistness cloths thermal efficiency (g_{cl},W)		0.122
Oxygen uptake (VO_2 max %)	%	30-40

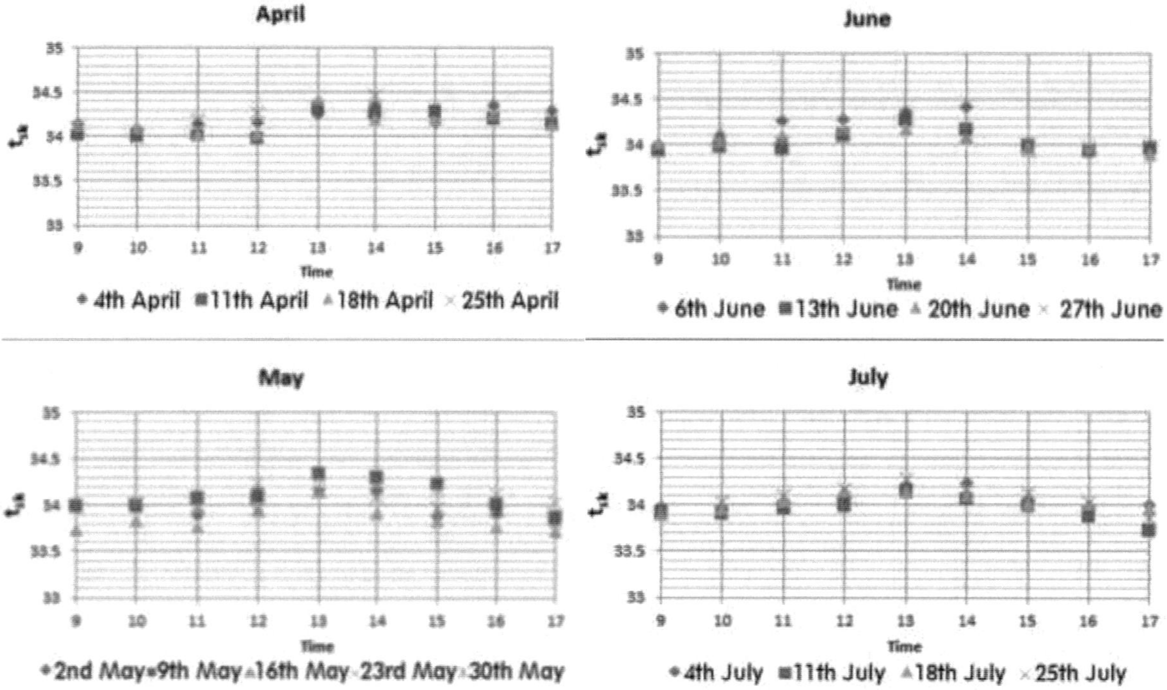

Figure 11. Predicted skin temperature, t_{sk} (April- July)

adjacent skin and creates scope for heat loss due to convection way. Figure 9 shows that air flow varies between 0.2 -0.8 ms^{-1}.

Figure 10 indicates the heat gain from radiation and indoor artificial light. It is around 600-800 lux. It represents the indoor light intensity during the class time. Heat gain occurs for the lighting intensity. These are the basic thermal parameters which are directly related to the heat loss and heat gain process. Data were collected by the pre-mentioned measuring tools and plotted by hourly basis. Every horizontal line indicates the data through the day mentioned in the figures (Figure 7-10).

3.2. PHS prediction calculation

For establishing PHS criteria, here established formulas was in term according to ISO standards 7933. For calculation of PHS, these formulas have been conducted.

For clothed subject skin temperature:

$$t_{sk} = 12.17 + 0.020\, t_a + 0.044\, t_r + 0.194\, P_a - 0.253\, V_a + 0.003\, M + 0.513\, t_{re} \quad (3)$$

Here for light work like studying, sitting metabolic rate was considered as 1.1 met = 58.2 W/m^2.
Following equations also considered for evaluating the PHS criteria.
[$t_{cr} = t_{crn}$, $HR = HR_n$]
For moderate heat stress range,

$$t_{cr} = t_{crn} + k_b (t_{sk} - t_{skp}) \quad (4)$$

$$HR = HR_n + k_{b, HR} (t_{sk} - t_{skp}) (5)$$

And for high heat stress range,

$$t_{cr} = t_{crn} + (C_k - t_{skp}) \quad (6)$$

$$HR = HR_n + k_{c, HR} (t_{sk} - t_{skr}), \text{ where } t_{skr} = 36^\circ C \ (7)$$

And for t_{sk} and t_{cr} the acceptance range is,
Everyday life body surface temperature, $t_{sk} = 34^\circ$C.
In the case of body temperature), $t_{cr} = 37.6 - 38^\circ$C.
For the evaluation of PHS, student's height, weight, body surface, metabolic rate, clothing and age are considered as per Table 4. From the study, it is found that during the month of April to July indoor AT become very high near about 34.5°C and RH about 75-85%. During this time period t_{sk} leads to higher than the regular comfortable skin temperature. Figure 11 represents the skin temperature throughout the research work. Data are plotted on hourly basis, and it shows that almost every time the temperature is above 34°C which indicates an uncomfortable climatic condition. Figure 12 shows that body surface temperature goes high about 61% time during this period of time because of high RH it creates an unfamiliar environment. Mostly at morning and noon time observed data shows that it is above the comfort range. Further t_{cr} became higher than the normal 38°C for high AT at indoor climate.

4. Conclusions

The major observation of this study is that indoor thermal parameters in classrooms are not satisfying during summer time. Environmental parameters influence a lot on thermal comfort and create health issues. In this study, heat stress criteria were observed in April-July, during the regular class time and from that major findings can be drawn like that:

- At Chittagong region indoor AT remain 30-34°C, while RH is around75-85% during operative time.
- Minimum ventilation, elevated temperature and high humidity reanimated heat gain from outdoor to indoor environment. As a result, heat gain happed to the occupants. So core body temperature and skin temperature of occupants exceed from its comfort range.
- During this summer time, PHS indicates health risk alert, shows elevated skin temperature during middle of the day (12:00 to 15:00) from the normal skin temperature (34°C).

Reference

[1] University Grants Commission of Bangladesh, access : http://www.ugc.gov.bd/en (01/01/2017)
[2] Astolfi A, Corgnatiand S P, Lo Verso V R M 2003 *Proc. Int. Conf. on Research in Building Physics, Leuven, Belgium*
[3] Nico M A, Liuzzi S and Stefanizzi P 2015 *Applied Ergonomics* **48** 111–120
[4] Ansaldi R, Corgnati S P and Filippi M 2007 *Proc. of Clima, Well Being Indoors* 3–7
[5] Kjellstorm T, Briggs D, Freyberg C, Lemke B, Otto M and Hyatt O 2016 *Annual Review of public Health* **37** 97-112
[6] Wong N H and Khoo S S 2003 *Energy and Buildings* **35** 337-51
[7] Corgnati S P, Filippi M and Viazzo S 2007 *Building and Environment* **42** 951-59
[8] Humphreys M A and Nicol J F 1998 *ASHRAE Transactions* **104** 991–1004
[9] Hwang R L, Cheng M J, Lin T P and Ho M C 2009 *Building and Environment* **44** 1128-34
[10] Newsham G R 1997 *Energy and Buildings* **26** 283-91
[11] Fong M L, Lin Z, Fong K F, Chow T T and Yao T 2011 *Indoor Air* **21** 231-39
[12] Haldane J S 1905 *Epidemiology & Infection* **5** 494–513

[13] Rowell L B 2011 *Comprehensive Physiology (Wiley Online Library)*
[14] Kjellstrom T, Holmer I and Lemke B 2009 *Global Health Action* **2.1** 2047
[15] A. S. H. R. A. E. Standard 55-2004 *Thermal environmental conditions for human occupancy* 9-11
[16] Malchaire J B M 2006 *Industrial Health* **44** 380-87

Nomenclature

Symbol	Meaning	Unit
AT	Air temperature	(°C)
f_{cl}	Clothing factor	
HR	Heart rate	(bpm)
HR_n	Equilibrium heart rate	(bpm)
I_{cl}	Clothing insulation	
k_b	Rate of increase in body temperature	(°C)
$k_{b/c,hr}$	Heart rate increase	(bpm/K)
M	Metabolic rate	(W/m^2)
MV	Mechanical ventilation	
NV	Natural ventilation	
P	Air pressure	(kPa)
PHS	Predicted heat strain	
p_a	Vapor pressure of air	(kPa)
R_{cl}	Clothing thermal insulation	
RH	Relative humidity	(%)
t_a,T_a	Air temperature	(°C)
t_{cl}	Surface temperature of clothing	(°C)
t_{cr}	Core temperature	(°C)
t_{crn}	Permanent body temperature	(°C)
t_r	Mean radiant temperature	(°C)
t_{re}	Rectal temperature	(°C)
t_{sk}	Skin temperature	(°C)
t_{skp}	Upper limit skin temperature	(°C)
v	Air velocity	(ms^{-1})
v_a	Volume of air	(m^3)

IOP Conf. Series: Journal of Physics: Conf. Series **1086** (2018) 012008 doi:10.1088/1742-6596/1086/1/012008

A comparison of the absorbed power for periodic, disordered and deterministic aperiodic arrays of Silver nanoparticles

S Sarwar[1], S Islam[1], T Hosain[2], H Sarwar[3] and M J Rashid[1]

[1] Dept. of Electrical and Electronic Engineering, University of Dhaka, Dhaka
[2] Dept. of Electrical and Electronic Engineering, University of Asia Pacific, Dhaka
[3] Dept. of Computer Science and Engineering, United International University, Dhaka

E-mail: sakin.sarwar1993@gmail.com

Abstract. The comparison between the optical properties of ordered and disordered metallic nanoparticles (NP) on the top of a silicon substrate has been studied. Such structures have been used to enhance the performance of optoelectronic devices, by exploiting the properties of the metallic NPs and their interaction with each other. This comparative study has been carried out using the Finite Difference Time Domain (FDTD) numerical analysis. Here, we report the role of distributed NP's on a silicon substrate and the electric field coupling between them. A slab of silicon substrate with a size of the micrometer range has been taken. Several metallic NPs have been distributed on the surface of the silicon. This structure has been studied for both periodic and disordered arrays of metallic nanoparticles. These NP's have been chosen as silver for which they exhibit strong localized surface plasmon resonance (LSPR). Total absorbed power has also been studied. The structure is illuminated by a plane wave source with a wavelength range that varies between 300 nm to 800 nm.

1. Introduction

Plasmonics is a new branch of research, which can be engineered into novel photonic and optoelectronic devices [1-6]. The unique electronic and optical properties of plasmonic nanoparticles (NP), specifically the localized surface plasmon resonance (LSPR) phenomenon are the core of plasmonic research [7-10]. The LSPR of the silver (Ag) NPs emerges from the interaction between the joint oscillation of conduction band electrons and incident electromagnetic radiations confined to the nanoscale [11]. LSPR extinction is a maximum when the frequency of the incident light and the free in the nanoparticles are matched [12,13].

The light matter interaction discussed, provides various results depending on the orientation of the NP medium. Ordered nanoparticles show distinct resonances, which can be controlled by the period of nanoparticles and spectrally red-shifts due to the phase alignment of scattered fields between neighboring particles [14]. If we consider deterministic aperiodic (DA) plasmonic structures, generated by mathematical sequences and number theory [15-19], it provides unique light localization and transport properties connected with the complex aperiodic structures [20-22].

In this work, a comparative study of the periodically and a periodically oriented nanoparticles have been performed using the Finite Difference Time Domain (FDTD) method. The discussion will be focused on the power absorption of the plasmonic nanostructure and how it is affected by size, periodicity and the DA arrays. Random structures have also been considered, which provide a whole different result than the periodic or DA arrays [23]. The analysis will be focused on the power absorption of DA arrays, produced with respect to the Thue -Morse sequence.

2. Simulation Methodology

Figure1 shows the proposed plasmonic nanostructure. The Ag NPs are arranged periodically over a glass substrate. The nanostructure is illuminated by a plane wave source of the wavelength range 200nm – 800nm.

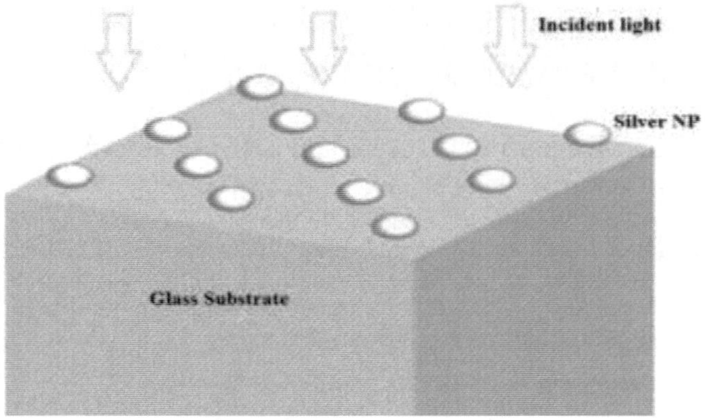

Figure 1. The layout of the simulation design

For such sub-wavelength metallic NPs, the interaction between them affects the absorption of light. Due to their plasmonic properties, multiple resonant modes are observed. For the nanostructure to behave as a photodetector, single resonant modes should be expected as results. To get single resonant modes, the periodicity of the Ag NPs must not be maintained, instead random arrays (figure 2b) or deterministic aperiodic (DA) (figure 2c) arrays can be used. This optimization of Ag NP distribution significantly changes the power absorbed. The nanostructure is simulated using the lumerical FDTD solutions software [24], a commercial electromagnetic solver based on the FDTD method. As mentioned above, both periodic and random nanostructures have been considered.

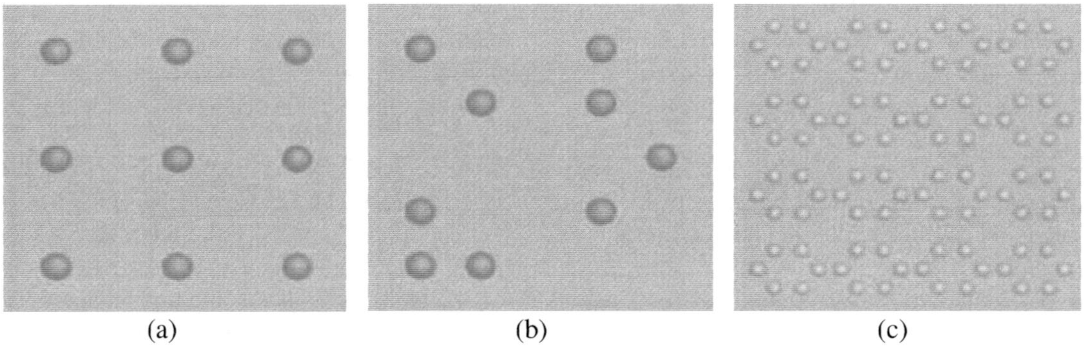

Figure 2. (a) Periodic arrangement of Ag NPs, (b) Random arrangement of Ag NPs and (c) Deterministic aperiodic array Ag NPs mainly following the Thue-Morse mathematical sequence.

The boundary conditions for the FDTD region are anti-symmetric for x-axis and symmetric for y-axis. The z-axis has the perfectly matched layer as boundary condition. The non-uniform meshes are applied in the entire simulation domain. The mesh size in the regions around the NPs is $\Delta x = \Delta y = \Delta z = 2\ nm$ and the mesh size in the silicon region is $\Delta x = \Delta y = 1\ nm$ and $\Delta z = 2\ nm$. The incident plane wave propagates from up to down in the z-direction. The transmission of optical energy is defined as the light power flowing into the substrate normalized by the incident light power. Ag has been used as the metallic NP since they are widely recognized for plasmonic research.

3. Results and Discussion

Figure 3 shows the effect of Ag NP diameter on the absorbed power of the nanostructure. The increase of the diameter leads to the formation of multiple resonant mode. For NPs smaller than incident wavelength, electromagnetic field is uniform across a particle such that the conduction electrons move in phase producing only dipole-type oscillations, resulting in single narrow surface plasmon resonance (SPR) peaks. As the size of the NP increases the gap between them decreases. Several resonant modes are now created at longer wavelengths. The field across each particle becomes non-uniform and this phase retardation broadens the dipole resonance, exciting higher multiple resonance.

Figure 4 shows the change in absorbed power when the NP orientation is random. The curves have only a single resonant peak from 50nm to 125nm NP diameter. Since energy of random systems can be discrete, disordered induced localized states appear at discrete resonant frequency, where the electronic or optical diffusive transport breaks down. For 150nm NP diameter, the increase in size decreases the gap between the randomly oriented NPs. The nanostructure now starts behaving like the periodically oriented one and creates a secondary resonant mode.

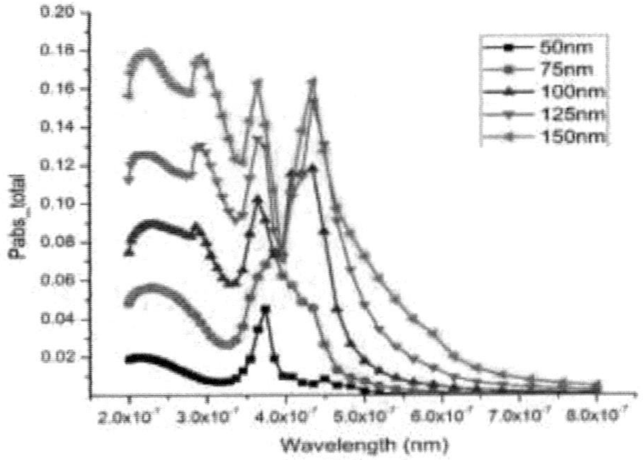

Figure 3. The total absorbed power versus wavelength study of periodically arranged Ag NPs on glass substrate for NP diameter ranging from 50nm to 150nm.

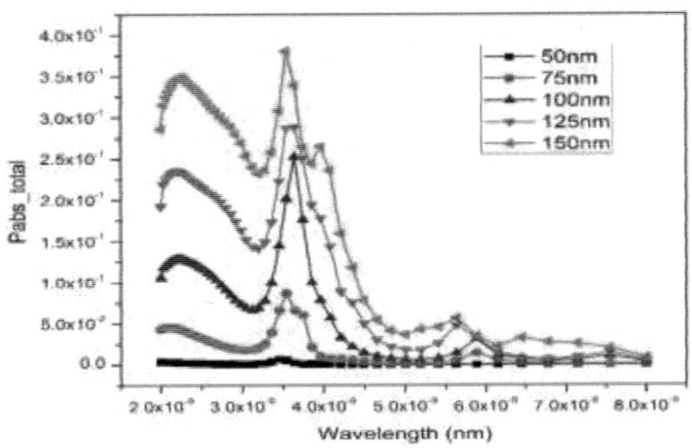

Figure 4. The total absorbed power versus wavelength study of randomly arranged Ag NPs on glass substrate for NP diameter ranging from 50nm to 150nm.

The Thue Morse sequence has been used as a deterministic aperiodic system and was found to provide the best result for the absorbed power (figure 5). Though it has shown some small resonant modes in the lower wavelength region, the total absorbed power had the greatest peak in the 350 nm region compared to the periodic and disordered arrays of NPs, resulting in lower power losses.

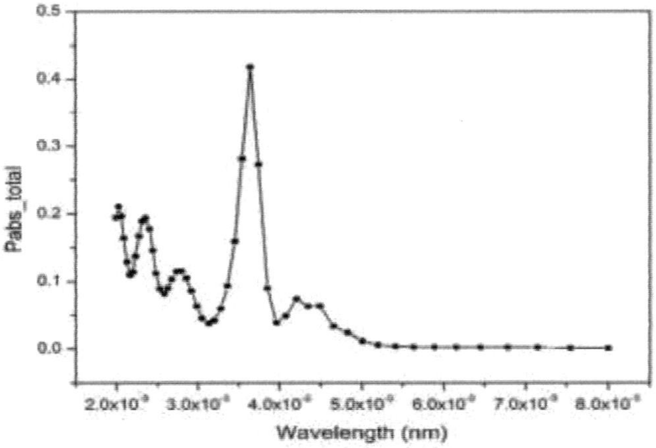

Figure 5. The total absorbed power versus wavelength study of Thue Morse array of Ag NPs on glass substrate for NP diameter of 50nm.

4. Conclusions

Comparing ordered NP's distribution with disordered ones, it is seen to produce less resonant modes for the latter one. The disordered array of NP's resembles the diffuse diffraction spectra of amorphous structures, that is, resulting in single resonant modes. Only for a diameter of 100 nm for the silver NP's, a second resonant mode of lower absorbed power than the primary resonant mode has been observed in the higher wavelength region. The ordered plasmonic nanostructure is thus ideal for use in photovoltaic applications whereas the random or deterministic aperiodic plasmonic nanostructure can be put into application as UV detectors due to peaks formed around the UV range.

References

[1] Berry C W, Wang N, Hashemi M R, Unlu M, and Jarrahi M 2013 *Nature Communications* **4** 1622
[2] Politano A 2012 *Philosophical. Magazine* **92 (6)** 768–778
[3] Bao Q and Loh K P 2012 *ACS Nano* **6 (5)** 3677–3694
[4] Politano A and Chiarello G 2013 *Nanoscale* **5** 8215–8220
[5] Politano A 2012 *Plasmonics* **7 (1)** 131–136
[6] Lakowicz J R 2006 *Plasmonics* **1 (1)** 5–33
[7] Wang J and Zhou H S 2008 *Analytical Chemistry* **80** 7174-78
[8] Jin Y and Gao X 2009 *Nature Nanotechnology* **4** 571-76
[9] Lal S, Link S and Halas N J 2007 *Nature Photonics* **1** 641-48
[10] Hiep H M, Yoshikawa H and Tamiya E 2010 *Analytical Chemistry* **82** 1221-27
[11] Xie T, Jinga C and Yi-Tao Long 2017 *Analyst* **142** 409-20
[12] Stewart M E, Anderton C R, Thompson L B, Maria J, Gray S K, Rogers J A and Nuzzo R G 2008 *Chemical Reviews* **108** 494-521
[13] Nehl C L and Hafner J H 2008 *Journal of Materials Chemistry* **18** 2415-19
[14] Lamprecht B, Schider G, Lechner R T, Ditlbacher H, Krenn J R, Leitner A and Aussenegg F R 2000 *Physical Review Letters* **84** 4721–24
[15] Queffelec M 1987 *Substitution dynamical systems-spectral analysis, Lecture Notes in Mathematics*, (Springer: Berlin, 1987) **1294** ISBN-978-3-540-18692-2 (Print) 978-3-540-48088-4 (Online)

[16] Macia E 2006 *Reports on Progress in Physics* **69** 397-441
[17] Williams S G 2002 *Symbolic dynamics and its applications* American Mathematical Society (Short Course, January 4-5, 2002, San Diego, California) **60**
[18] Schroeder M R 1985 *Number Theory in Science and Communication* (Springer-Verlag Berlin Heidelberg), ISBN 978-540-26598-6
[19] Prusinkiewicz P and Lindenmayer A 1990 *The Algorithmic Beauty of Plants* Springer, New York, 1990
[20] Janot C 1997 *Quasicrystals: A Primer* 2nd ed. Oxford University Press, New York, 1997
[21] Dulea M, Johansson M and Riklund R 1992 *Physical Review B* **45** 105–14
[22] Kroon L, Lennholm E and Riklund R 2002 *Physical Review B* **66** 094204
[23] Forestiere C, Miano G, Boriskina SV, Dal Negro L 2009 *Optics Express* **17(12)** 9648-61
[24] Lumerical Solutions, Inc. 3 *http://www.lumerical.com/*

IOP Conf. Series: Journal of Physics: Conf. Series **1086** (2018) 012009 doi:10.1088/1742-6596/1086/1/012009

Deposition and Optical Characterization of ZnO Thin films on Glass Substrate

S Islam[1], K M A Hussain[2] and M J Rashid[1]

[1]*Dept. of Electrical and Electronic Engineering, University of Dhaka, Bangladesh.*
[2]*Experimental Physics Division, Atomic Energy Centre, Bangladesh Atomic Energy Commission, 4 KaziNazrul Islam Avenue, Ramna, Dhaka.*

E-mail: islam.samia1992@gmail.com mjrashid@du.ac.bd

Abstract. ZnO is a wide band gap (3.37 eV) compound semiconductor which can be used in short wavelength optoelectronic devices.Here in this work we chose spin coating method as growth mechanism as it is simple, less expensive and in addition film uniformity is maintained in this method.In the growth process zinc acetate dihydrate, iso–propanol and monoethanolamine were used for producing Sol-gel Solution.The prepared solution was kept for 3 days (44 h) and 7 days (188h) for aging. After deposition, optical characterization has been performed by UV-VIS-Spectrometry The Transmittance values of the films deposited at the speeds of 2500 rpm (with 44 h aged solution) and 3500 rpm (with 188h aged solution) are at the range of 70-73% and the Energy band gaps of the films are 3.1 eV and 3.25eV, respectively. Whereas films deposited at the speed of 4000 rpm shows 86%transmittance and band gap energy of 3.3 eV which is much closer to the optimum value of 3.37 eV. Lower absorption values have been shown for higher coating speed.So with the increase in coating speed,the transmittance and the band gap both are increased. Also we can see that there is very negligible effect of aging time on optical properties of deposited films.

1. Introduction:

Transparent conducting oxide layers have been studied widely because of their far-reaching range of applications. Recently, ZnO thin films have urged a great interest as transparent conductor, owing to their favorable properties.ZnOis a key technological material which has wide band-gap energy of 3.37 eV[1]. Advantages associated with a large band gap include higher breakdown voltages [2], ability to sustain large electric fields [3], lower electronic noise [2], high-temperature and high-power operation [4].

ZnO thin films are low cost, nontoxic and highly durable against hydrogen plasma compared to ITO [5]. ZnO-based nanostructures including nanowire arrays hold a host of opportunities for flat screen displays, field emission sources, gas, chemical [6] and biological sensors, and as UV light emitters and switches [6-8].The high exciton binding energy (60meV) [1] in ZnO crystal can ensure efficient excitonic emission at room temperature. Furthermore, these nanostructures are easily formed even on cheap substrates such as glass and hence they have auspicious potential in the nanotechnology future.

ZnO films have been fabricated using various kinds of deposition techniques. Among them molecular beam epitaxy, chemical vapor deposition, chemical bath deposition and spin coating are prevalent. It seems that the vacuum based technology is costly which might not be suitable for many applications. Thus low cost techniques such as chemical bath deposition and spin coating are

IOP Conf. Series: Journal of Physics: Conf. Series **1086** (2018) 012009 doi:10.1088/1742-6596/1086/1/012009

promising for different applications. In our study we have chosen Sol- gel Spin Coating as our deposition method because of controllability of composition, simplicity in processing, cost effectiveness and Film Uniformity. Moreover this process has been widely used in the manufacture of integrated circuits [9], optical mirrors, color television screens and magnetic disk for data storage [10].

In this paper we provide a meticulous overview on optical properties of ZnO nanostructures. Production of high quality films at relatively lower annealing temperature, optical characterization of ZnO deposited films to understand their optical properties and investigation of the optimum film on the basis of characterization results are the prime concern here. The assessment is organized as follows: At first we discussed a complete review of sample grounding, sol preparation and fabrication of films of ZnO. Next optical characterization results anddiscussion on optical properties arepresented. Finally some conclusion and aviewpoint for the future are given.

2. Experimental Method:

For preparing sol, we have used Zinc acetate dehydrate $Zn(CH_3COO)_2.2H_2O$ (ZAD) as the precursor material and chosen the concentration of the solution 0.5 M. In order to produce the solution of required concentration, we have taken 4.39 g of Zinc Acetate Dehydrate. We weighted the required amount of $Zn(CH_3COO)_2$ using a balancing instrument. 4.39 g of Zinc acetate dehydrate was mixed with 100 ml of Iso- propanol. Then we started boiling the prepared solution at 65 °C under constant stirring using a magnetic stirrer. The color of the solution turns milky after 15 minutes of stirring and heating. After 30 minutes of continuous stirring and heating, we have added 6-10 drops of monoethanolamine to the solution by a 5 ml dropper. Monoethanolamine acted as sol stabilizer. The solution turned transparent, but it's heating and stirring was continued for further 90 minutes. Then the solution under stirring was allowed to cool. The whole process of Sol preparation took 2 hours and we have used CBD machine for preparing this Solution of Deposition.

Aging mechanisms have a great influence on the formation of good quality ZnO nanostructure. So we have decided to vary the aging time to see its effect on the formation of the films. We have kept the solution for 44 hours (almost 3 days) and 188 hours (almost 7 days) at room temperature to produce aged solution. We haven't prepared any film from freshly prepared solution.

Once the fabrication has been done, the fabricated films were taken to the vacuum heat chamber for annealing treatment. Here, a pre-heat treatment of 200°C was applied for 1 hour. Post heat treatment temperature was 250°C and the duration of the treatment was 1 hour.

3. Results and Discussion:

After the preparation of the samples, the optical properties were investigated. We have fabricated eight Samples varying aging time and coating speed. Among them we have chosen four samples for characterization.

We have listed the samples in the table 1 that are being characterized-

Table 1 List of prepared Samples

Sample ID	Aging time	Coating Speed
S1	44 hours	2500 rpm
S2	44 hours	4000 rpm
S3	188 hours	3500 rpm
S4	188 hours	4000 rpm

(a) Transmittance Analysis:

Transmittance of the surface of a thin film is its effectiveness in transmitting radiant energy. Fig 1 (a) and (b) shows the variation of transmittance with respect to wavelength. These showed variations are for films of different speeds produced by 44 h and 188 h aged solutions.

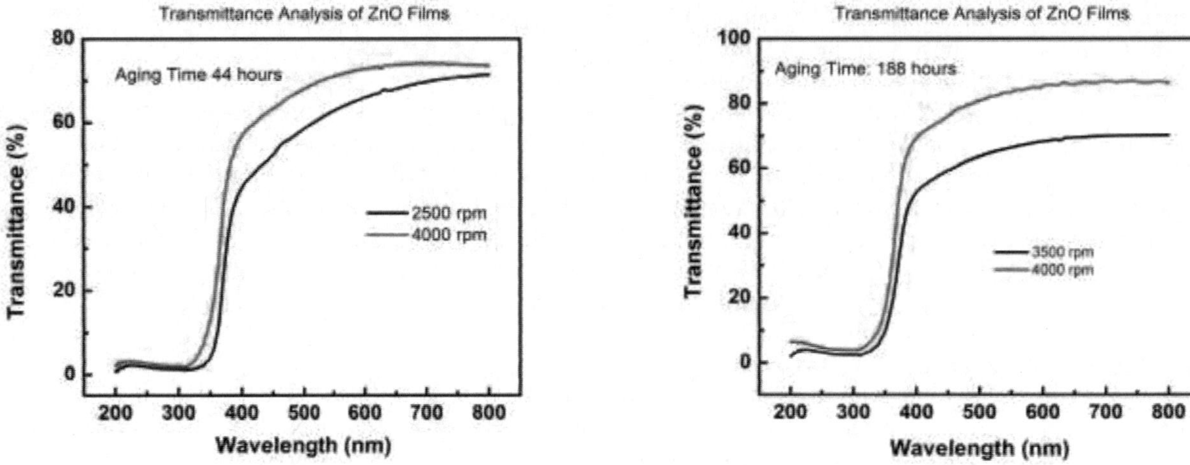

(a) (b)

Figure 1. Transmittance Analysis of fabricated ZnO films

In Fig 1 (a), it is seen that films with different rpm speed (aging Time 44 h) have almost same transmittance pattern. It is also observed that transmittance rises very rapidly after 350 nm, attain at maximum value and then stay in between 70-75% up to 800 nm. When the coating speed is 2500 rpm, the maximum transmittance value is70% whereas at the speed of 4000 rpm the value is reported 75%. Similarly if we look at Fig 1(b), we can observe that in case of 3500 rpm coating speed, the value of transmittance is reported 76% but at the speed of 4000 rpm a dramatic increase in transmittance value has been observed and that was up to 87%.

(b) Absorbance Analysis:

The term absorption refers to the physical process of absorbing light, while absorbance does not always measure absorption: it measures attenuation (of transmitted radiant power).Fig 2(a) and (b) shows the absorbance spectra of ZnO thin films (different aging time) with respect to wavelength-

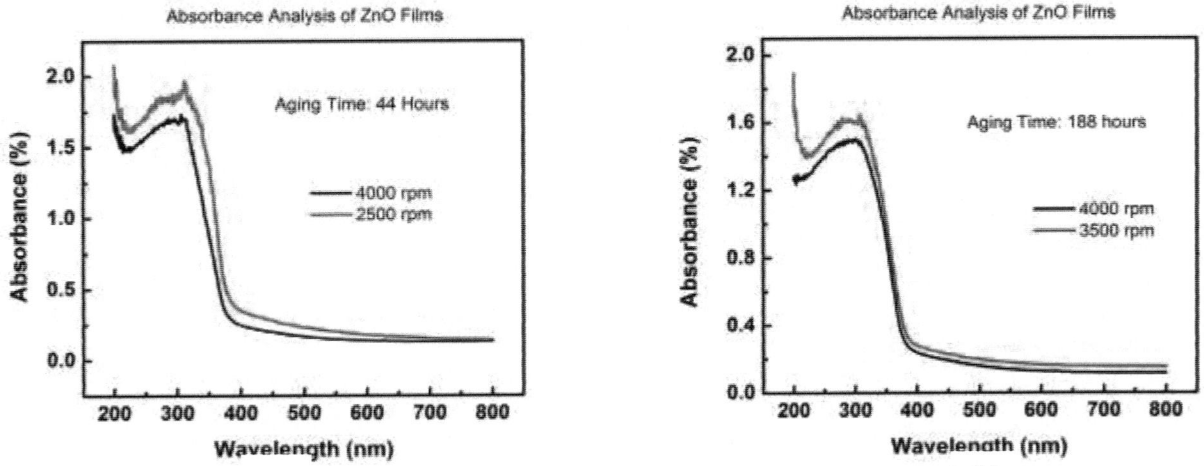

(a) (b)

Figure 2. Absorbance Analysis of fabricated ZnO films

If we look at Fig 2 (a) and (b) we can see that, absorbance value rises very rapidly before 300 nm, attains at maximum value and then decreases rapidly up to 380nm. It is also observed that in case of films produced by 44 h aged solution the absorbance range is in between (0.3~1.8) and for the films produced by 188 h aged solution, the absorbance ranges in between (0.2~1.6). From Fig 4.8 and 4.9, it is clear to us that, with the increase in coating speed, absorbance decreases.

(c) Band Gap Measurement:

In Fig 3 (a) and (b), $(\alpha h\nu)^2$as a function of photon energy is plotted to obtain band gap energy for different ZnO films. Table 2 summarizes the band gap of different films-

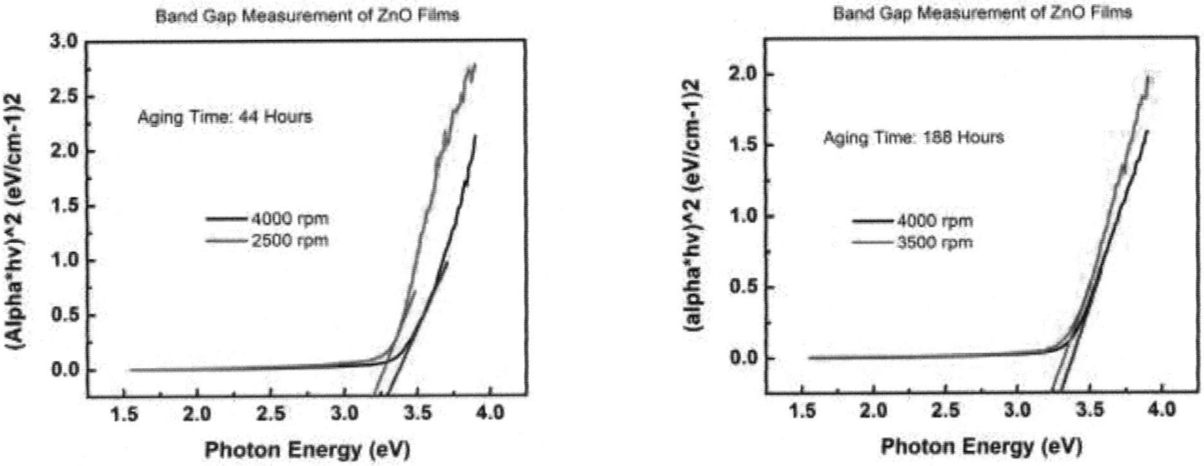

(a) (b)

Figure 3. $(\alpha h\nu)^2$ vs. photon energy curves for ZnO thin films

Table 2 Band Gap of different ZnO thin films

Sample ID	Coating Speed	Aging Time	Band gap (eV)
S1	2500 rpm	44 h	3.2
S2	4000 rpm	44 h	3.3
S3	3500 rpm	188 h	3.25
S4	4000 rpm	188 h	3.3

If we look at the band gap values, it has been observed that the plots of $(\alpha h\upsilon)^2$ versus $h\upsilon$ are linear over a wide range of photon energies. In case of films produced by 44 h aged solution, we can see that at 2500 rpm speed the obtainable band gap value is 3.2 eV. On the other hand when the speed increases up to 4000 rpm the band gap value increases to 3.3 eV which is much closer to the optimum value 3.37 eV. Same thing has been observed in films when aging time is 188 h. The band gap value improves as we increase the coating speed. At 4000 rpm coating speed, 3.3 eV band gap has been observed which matches the actual band gap value.

We can also evaluate from the above table that, the changes in the value of band gap energies with the change in aging time is very much negligible. With the same speed of 4000 rpm, we have got band gap energy of 3.3 eV regardless of aging time. That means whether we take 44 h aged solution or 188 h aged solution, prepared films will give almost same optical band gap unless we change the coating speed.

4. Conclusions

Investigating the effects of coating speed and aging time on diverse properties of ZnO thin films and hence finding the optimized result is the prime concern of our work. The Transmittance values of the films deposited at the speeds of 2500 rpm (with 44 h aged solution) and 3500 rpm (with 188h aged solution) are at the range of 70-73% and the Energy band gaps of the films are 3.1 and 3.25 respectively. Whereas films deposited at the speed of 4000 rpm shows 86% transmittance and band gap energy of 3.3 eV. So After scrutinizing all the characterization results, it is clear to all that film quality greatly depends on coating speed. On the other hand aging time hardly affects the quality of the fabricated films. Films deposited at 4000 rpm exhibit higher quality and could be used in short wavelength optoelectronic devices.

Acknowledgements
This research work was done at Experimental Physics Division of Bangladesh Atomic Energy Commission.

References:

[1] Gao P X and Wang Z L 2004 *J. Phys. Chem. B***108**7534
[2] Corso A D, Posternak M, Resta R, and BaldereschiA 1994 *Physical Review* **50** 10715
[3] Özgür Ü, Alivov Y I, Liu C, Teke A, Reshchikov M A, Doğan S, Avrutin V, Cho SJ and Morkoç H 2005 *J. Appl. Phys.* **98** 041301
[4] NASA Scientific and Technical Information 2006 Retrieved 17 October 2009. JPL scientists developed UV-protective sunglasses using dyes and "zinc oxide, which absorbs ultraviolet light".
[5] Brattain W H, entry of 15 December 1947, laboratory notebook, case 38139-7. Bell Laboratories archives.
[6] NiedduM, Boatto G, Pirisi M A and Dessì G 2010 *Rapid Commun. Mass Spectrom.***24** 2357
[7] Padmavathy N and Vijayaraghavan R 2008 *Sci. Technol. Adv. Mater.* **9** 035004
[8] Hughes G, McLean N R 1988 *Arch Emerg Med.* **5** 223
[9] Mihi A, Ocaña M, and Míguez H 2006*Adv. Mat.* **18**2244
[10] Burroughes J H, Bradley D D C, Brown A R, Marks R N, Mackay K, Friend R H, Burns P L and Holmes A B1990 *Nature* **347** 539

IOP Conf. Series: Journal of Physics: Conf. Series **1086** (2018) 012010 doi:10.1088/1742-6596/1086/1/012010

Experimental study of the optical properties of Cu_2ZnSnS_4 thin film absorber layer for solar cell application

S I Swati[1], R Matin[2], S Bashar[2] and Z H Mahmood[1]

[1] *Dept. of Electrical and Electronic Engineering, University of Dhaka, Dhaka 1000, Bangladesh.*
[2] Institute of Fuel Research and Development (IFRD), *Bangladesh Council of Scientific and Industrial Research (SCSIR), Dhaka, Bangladesh.*

E-mail: supriya.swatidu@gmail.com , zahid@du.ac.bd

Abstract. Cu_2ZnSnS_4 (CZTS) has become the subject of considerable interest due to its use as thin film solar cell material. Current thin film solar technologies based on $CuIn_xGa_{1-x}Se_2$ (CIGS) and CdTe photo-absorber materials use rare and expensive elements, such as In, Te, Ga and toxic Cd which severely limit the mass productions and deployment of solar cells. CZTS is an intrinsically p-type material with a large optical absorption coefficient ($\alpha = 10^4$ cm^{-1}) and exhibits a tunable direct optical band gap in the range between 1.49 to1.51 eV. The CZTS thin films were deposited on soda lime glass by dip coating method with sulfurized Cu, Zn, Sn precursors and were characterized CZTS thin film as absorber layer of solar cell. The optical characterization has been performed by UV-Vis spectrometry in order to study the effect of Cu/(Zn+Sn) ratio on the absorption coefficient and the band gap energy of the prepared thin films. As the ratio is increased (0.4 to 0.7), the absorption coefficient increases slightly whereas the band gap energy decreases.

1. Introduction

The ability to combine cheap abundant materials with low cost and low power deposition techniques is attractive for large scale and sustainable Photovoltaic. The prime candidate is Cu_2ZnSnS_4 (CZTS), an abundant, cheap and non-toxic alternative of traditional solar cell elements. CZTS is an I_2-II-IV-VI_4 quaternary compound. From the chalcopyrite CIGS structure, CZTS can be obtained by substituting the trivalent In/Ga with a bivalent Zn and IV-valent Sn which forms in the kesterite structure [1]. CZTS is an intrinsically p-type material. Its large optical absorption coefficient ($\alpha = 10^4$ cm^{-1}) [2] and tunable direct optical band gap in the range of 1.49-1.51 eV [3], make it attractive as absorber layer material.

Several vacuum deposition processes, such as sputtering [4], co-evaporation [5] and pulsed laser deposition [6], have been studied for CZTS thin-film preparation. However, these processes usually use a high-cost, precise vacuum apparatus and technology. Chemical bath deposition processes, such as electroplating [7], spin coating [8], and spray pyrolysis [9], also have been applied as CZTS thin-film preparation processes instead of vacuum deposition process. Metal hydrates are often employed as low-cost raw materials for the chemical bath deposition. Katagari and coworkers have fabricated CZTS solar cells with the CZTS absorber layer prepared by sulfurizing electron-beam evaporated precursors [10]. The efficiency of the solar cell was 7%. The solar cell consisted of an Al/ZnO:Al/CdS/CZTS/Mo/soda-lime glass (substrate) structure. The estimated optical band gap energy was 1.51 eV. Ito and Nakazawa reported on the fabrication of CZTS thin films prepared by atomic beam sputtering [11]. Friedlmeirer et al deposited thin films by thermal evaporation [12]. Tanaka et al. reported on CZTS thin film fabrication prepared by co-evaporation [13]. Seol et al. prepared CZTS thin films by RF magnetron sputtering [14]. The examined optical band gap energy was 1.51 eV. However all these processes were done in vacuum and they were expensive.

IOP Conf. Series: Journal of Physics: Conf. Series **1086** (2018) 012010 doi:10.1088/1742-6596/1086/1/012010

In this report, CZTS thin films were prepared by dip coating method with sulfurized Cu, Zn, Sn precursors and the influence of Cu/(Zn+Sn) ratio on the absorption coefficient and the band gap energy of the prepared thin films were studied.

2. Experiment

Copper (II) acetate monohydrate, zinc (II) acetate dihydrate and tin (II) chloride dihydrate and thiourea were dissolved in 2-methoxyethanol to prepare the Cu, Zn and Sn precursors. The chemicals were stirred by magnetic stirrer at 65^0C for 1 h to prepare the solution. Monoethanolamine was used as the stabilizer during the experiment. Soda Lime Glass substrates were first cleaned by ultrasonic bath for 45 minutes sequentially with methanol, acetone, de-ionized water and acetone. Then after drying the substrates they were dip coated in the solution for two or three minutes. After deposition by dip coating, all the films were dried at 250°C for 1 h to eliminate solvents. The coated glasses were sulfurized by annealing at 400°C in a hydrogen sulfide containing atmosphere for 1 h.

Four types of solution were prepared by varying the Cu/(Zn+Sn) ratio from 0.4 to 0.7. The ratio of Zn/Sn was kept constant (0.55). Table-1 lists the chemical composition of the prepared tin films. After fabrication, optical characterization has been performed by UV-VIS-Spectrometry to study the effect of Cu/(Zn+Sn) ratio on the absorption coefficient and band gap energy of the prepared thin films.

Table 1. Chemical compositions of sol-gel solutions and CZTS thin films

Solution Name	Sample Name	Cu/ (Zn+Sn)	Zn/Sn
Sol-1	CZTS-1	0.4	0.55
Sol-2	CZTS-2	0.5	0.55
Sol-3	CZTS-3	0.6	0.55
Sol-4	CZTS-4	0.7	0.55

3. Result and discussion

To investigate the optical properties of the prepared thin films, the UV-Vis spectroscopy analysis was done with wavelength ranging from 300 to 800 nm. The absorption coefficient, α was calculated from the measured absorbance data for different wavelength corresponding to different photon energy at room temperature. Fig. 1 shows the variation of α with the photon energy hν for different CZTS thin films. The absorption coefficient increases with photon energy but at high photon energy, it starts to decrease. Here, CZTS-3 has the highest absorption coefficient ($>1.5\times10^4$ cm^{-1}). CZTS-1 has the lowest absorption coefficient ($\sim10^4$ cm^{-1}). As the solution gets thicker the samples made from it achieves greater coefficient. CZTS-2, 3 and 4 have absorption coefficient greater than 10^4 cm^{-1} which is good for the absorber layer.

IOP Conf. Series: Journal of Physics: Conf. Series **1086** (2018) 012010 doi:10.1088/1742-6596/1086/1/012010

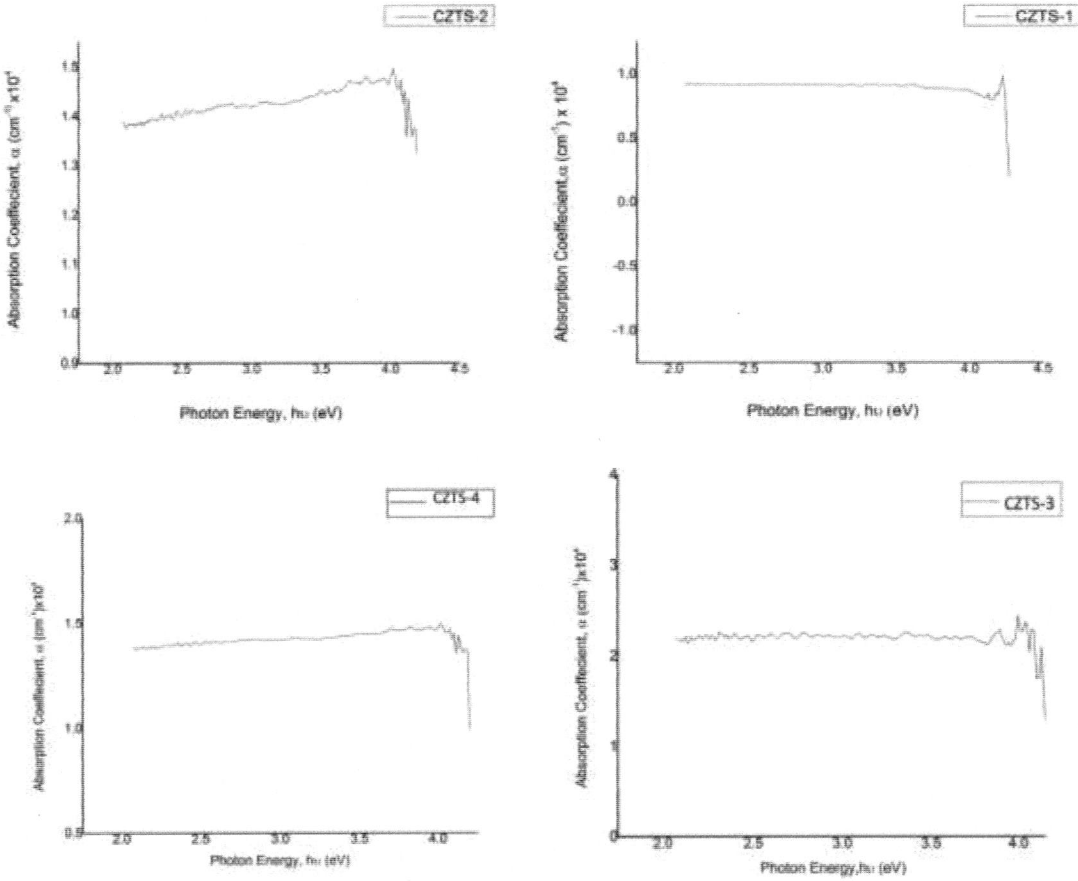

Figure 1. Absorption Coefficient vs. hv graph for CZTS thin films

The dependence of absorption coefficient on the photon energy helps to study the band structure and the type of electron transition involved in absorption process. The direct transition energy band gap can be obtained by plotting $(\alpha h\nu)^2$ vs. hν curve and then extrapolating the linear portion of the curve to $(\alpha h\nu)^2 = 0$. In Fig. 2 $(\alpha h\nu)^2$ as function of hν is plotted to obtain band gap for different CZTS thin films.

From figure 2 it is clearly seen that as the Cu to (Zn+Sn) ratio increases the exhibited band gap energy of the thin films decreases. Ideal band gap of CZTS film is 1.5 eV. CZTS-1 and CZTS-2 thin films has band gap much higher than the ideal one. CZTS-3 has band gap of 1.72 which is also higher than 1.5 eV. CZTS-4 has the band gap close the ideal one. So considering all the optical properties determined by UV-Vis spectroscopy CZTS-4 is the best choice as absorber layer in solar cell.

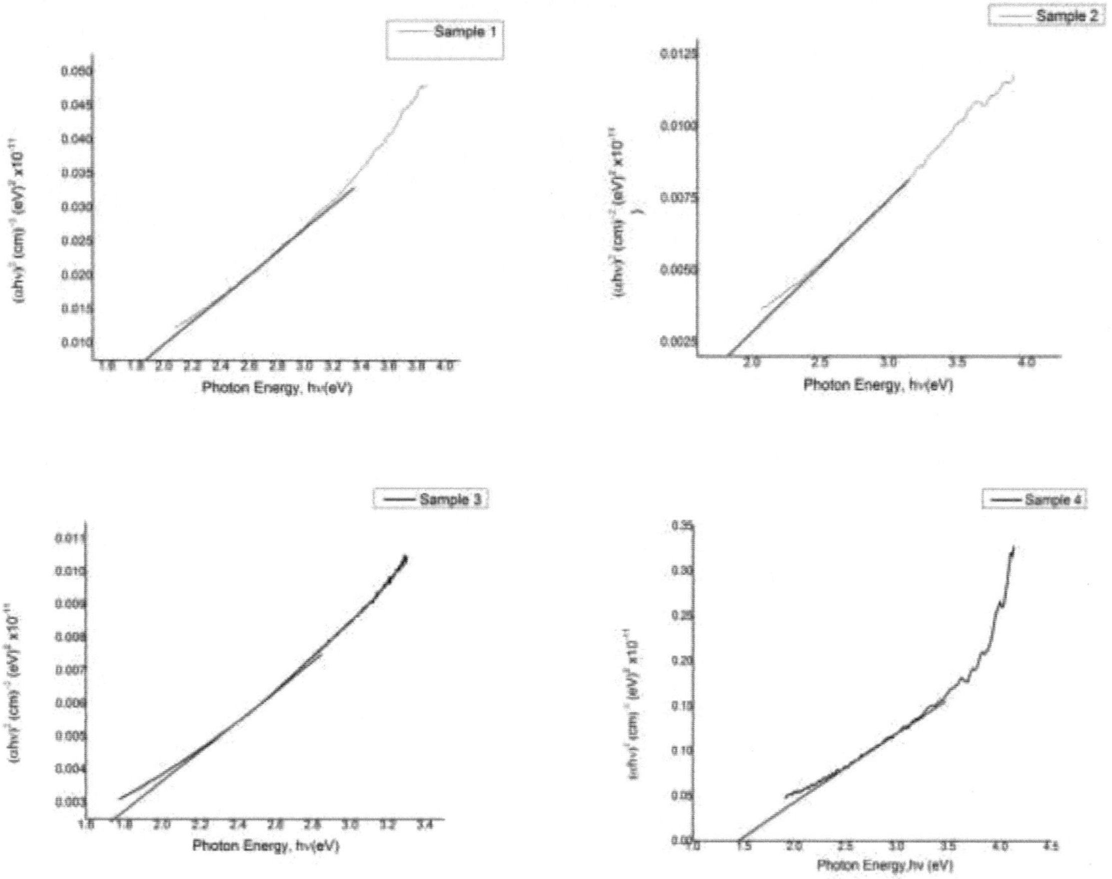

Figure 2. $(\alpha h\nu)^2$ vs. hν curve for CZTS thin films

4. Conclusion

CZTS thin film as absorber material in solar cell were fabricated from different concentrated metallic salt solutions with variation in the Cu/(Zn+Sn) ratio from 0.4 to 0.7. UV-Vis spectroscopy, confirms absorption coefficient of CZTS films in order of 10^4 cm^{-1}. As Cu/(Zn+Sn) ratio increases we find that the absorption coefficient increases slightly, whereas band gap decreases from 1.83 to 1.49 eV. The films could be applicable as solar absorber layers in CZTS thin-film solar cell application.

Acknowledgement

Authors are grateful to Institute of Fuel Research and Development (IFRD) of Bangladesh Council of Scientific and Industrial Research (BCSIR) for their experimental support during the research.

References

[1] Shockley W and Queisser H J 1961 *Journal of Applied Physics* **32** 510
[2] Tanaka K 2015 CZTS *Copper Zinc Tin Sulfide-Based Thin-Film Solar Cells (Wiley) pp-271*
[3] Olekseyuk I D, Gulay L D, Dydchak I V, Piskach L V, Parasyuk O V and Marchuk O V 2002 *Journal of Alloys and Compounds* **340 (1-2)** 141–145
[4] Amal M and Kim K H 2013 *Thin Solid Films* **534** 144
[5] Tanaka T, Kawasaki D, Nishio M, Guo Q and Ogawa H 2006 *Phys. Stat. Sol. (c)* **3** 2844
[6] Wibowo R A, Lee E S, Munir B and Kim K H 2007 *Phys. Stat. Sol. (a)* **204** 3373
[7] Gurav K V, Pawar S M, Shin S W, Suryawanshi M P, Agawane G L, Patil P S, Moon J H, Yun J H and Kim J H 2013 *Appl. Surf. Sci.* **283** 74
[8] Swami S K, Kumar A and Dutta V 2013 *Energy Procedia* **33** 198
[9] Khan M A M, Kumar S, Alhoshan M and Dwayyan A S A 2013 *Optics & Laser Tech.* **49** 196

IOP Conf. Series: Journal of Physics: Conf. Series **1086** (2018) 012010 doi:10.1088/1742-6596/1086/1/012010

[10] Katagiri H, Sasaguchi N, Hando S, Hoshino S, Ohashi J and Yokota T 1997 *Solar Energy Materials and Solar Cells* **49** 407–414

[11] Ito K and Nakazawa T 1988 *Japanese Journal of Applied Physics* **27** 2094-97

[12] Friedlmeier T M, Wieser N, Walter T, Dittrich H, Schock H W 1997 Proc. of the 14th European Conference of Photovoltaic Science and Engineering and Exhibition, Bedford. 1242.

[13] Tanaka K, Moritake N and Uchiki H 2007 *Solar Energy Materials and Solar Cells* **91** 1199-1201

[14] Seol J, Lee S, Lee J, Nam H, Kim K 2003 *Sol. Energy mater. Sol. Cells* **75** 155

IOP Conf. Series: Journal of Physics: Conf. Series **1086** (2018) 012011 doi:10.1088/1742-6596/1086/1/012011

Highly active carbon supported Sn/SnO_2 photocatalysts for degrading organic dyes

O R Das, M T Uddin, M M Rahman and M C Bhoumick

Department of Chemical Engineering and Polymer Science, Shahjalal University of Science and Technology, Sylhet 3114, Bangladesh

E-mail: mtuddin-cep@sust.edu; mtuddin_cep@yahoo.com

Abstract. In this work, the heterostructure carbon supported tin/tin oxide (Sn/SnO_2) photocatalysts has been developed. The target heterostructure was prepared by precipitation method followed by carbothermal reduction using tin tetrachloride ($SnCl_4$) as precursor for SnO_2, activated carbon as reducing agent, and ammonia solution as precipitating agent. The photocatalytic activity of the as-prepared photocatalysts was investigated by degrading methylene blue (MB) dye. The photocatalytic decomposition of methylene blue in solution under UV light irradiation revealed that the heterostructure Sn/SnO_2 photocatalysts catalyst led to higher activities than pure SnO_2. The suppression of charge separation in the heterostructure carbon supported Sn/SnO_2 due to space charge region between SnO_2 and Sn led to higher catalytic activity for the degradation of MB. The results also showed that the pH of solution had a direct influence on the photocatalysis process and neutral pH was favorable for the degradation of MB. In addition, the photocatalysts could be easily recycled without significant change in the catalytic activity which indicated the stability and reproducibility of the catalysts. The results suggested that the as-prepared carbon supported Sn/SnO_2 can be applied as a promising photocatalyst for the treatment of wastewater containing organic pollutants.

1. Introduction

Rapid growth in the industrial sector is playing a vital role in the economy of Bangladesh. Mainly the growth has been concentrated in garments which are export oriented industries. To support garments, a large number of other textile production processes including the washing, dyeing and finishing (WDF) of textiles and the number of wet processing units have been established and more are growing to be set up shortly [1]. A large quantity of water is used for dyeing and washing purposes. It is reported that 2 million m^3 effluents are discharged to water bodies every day from Textile sector in Bangladesh [2] which causes serious water pollution. This polluted water is causing serious health hazard in the neighbourhood, damaging fertility of the land, killing fishes and aquatic lives. Moreover, most industrial dyes are toxic, carcinogenic, and mutagenic, and have low biodegradability. Thus, finding a sustainable solution to treat the wastewater discharged from fast growing industry is of high priority. Currently available conventional water treatment methods such as adsorption, coagulation, sedimentation, filtration, chemical and membrane technologies involve high operating costs and could generate toxic secondary pollutants into the ecosystem [3]. In this regard, it is necessary to adopt reactive systems much more effective than those adopted in conventional purification processes. Therefore, the development of eco-friendly methods of destroying these pollutants became an imperative task.

IOP Conf. Series: Journal of Physics: Conf. Series **1086** (2018) 012011 doi:10.1088/1742-6596/1086/1/012011

Over the past decade, many research efforts have been devoted around the world to develop a newer, more powerful, and very promising technique called Advanced Oxidation Processes (AOPs) to treat the contaminants of drinking water and industrial effluents. Among these AOPs, heterogeneous photocatalysis employing semiconductor catalysts (TiO_2, ZnO, SnO_2, Fe_2O_3, CdS) has demonstrated its efficiency in degrading a wide range of ambiguous refractory organics into readily biodegradable compounds, and eventually mineralized them to innocuous carbon dioxide and water [4-8]. However, the photoinduced charge carrier in single bare semiconductor particles has a very short lifetime because of the high-recombination rate of the photogenerated electron/hole pairs, which reduces photocatalytic efficiency and hinders its further application in industry [9, 10]. Therefore, suppression of the recombination of photogenerated charge carriers in a semiconductor and effective utilization of solar energy still remain challenge in photocatalytic application. Among various techniques, deposition of metal on the semiconductor to reduce the charge recombination has received the most attention recently owing to their excellent catalytic activity.

Tin oxide (SnO_2), cassiterite structure, is a typical wide band gap n-type semiconductor (3.8 eV) [11]and one of the most widely used semiconductor oxides due to its chemical and mechanical stabilities. It is well known that Sn and SnO_2 nanomaterials can be used in energy production/storage devices. Herein, we report the synthesis of tin/tin oxide (Sn/SnO_2) nanocrystallites by novel carbothermal reduction method, while the activated carbon acts as both reducing agent and support in formation of $Sn\text{-}SnO_2/C$ catalyst. Photocatalytic properties of Sn/SnO_2 nanoparticles were studied through the degradation and mineralization of methylene blue in water upon illumination with UV light. This simple technique to synthesize Sn/SnO_2 nanoparticles could allow its wide-scale use for the degradation of methylene blue and other toxic organic compounds

2. Materials and Methods

2.1. Preparation of photocatalyst

The heterostructured Sn/SnO_2 nano-photocatalyst was prepared by precipitation method followed by carbothermal reduction method using tin tetrachloride ($SnCl_4$) as precursor and 12.5% Ammonia solution (NH_4OH) as precipitating agent and activated carbon as reducing agent. In a typical procedure, 5 mL of $SnCl_4$ was dissolved in 100 mL distilled water in an ice bath. Then 1g activated carbon prepared from peanut shell was added to the $SnCl_4$ solution followed by drop wise addition of ammonia solution under continuous stirring. A white precipitate of tin hydroxide [$Sn(OH)_4$] was appeared with the addition of ammonia solution. The addition of ammonia solution was continued until the pH of the solution was neutral. Subsequently, the resulting was kept stirring overnight. The mixture was then separated by centrifuging at 4000 rpm and washed several times with distilled water. The wet powder was then dried at 100 ^{0}C for 10 hour followed by calcination at 800 ^{0}C for 2 hour. Similar procedure was followed for the preparation of pure SnO_2 except that no carbon was added to $SnCl_4$ solution before adding ammonia solution. The synthesis of photocatalyst was optimized by changing the calcination temperature and time.

2.2. Evaluation of photocatalytic activity

Photocatalytic activities of the as-synthesized samples were evaluated by the degradation of methylene blue (MB) dye. All the experiments were conducted at room temperature open to air. In most of the experiment, 0.1 g photocatalyst was dispersed in 200 mL of MB (10 mg/L) to obtain the concentration of the catalyst at 0.50 g/L. The experiments were carried out in a pyrex beaker illuminated with a 125 W high pressure mercury lamp. The UV lamp was positioned above the solution beaker. Prior to irradiation, the suspension was stirred in the dark for 30 min to reach adsorption/desorption equilibrium. The solutions were continuously stirred during the experiments. At given irradiation time intervals, 5 mL of the suspensions were collected, and then centrifuged (4000 rpm, 10 min) to separate the photocatalyst particles. The concentrations of MB were measured by UV-Visible

IOP Conf. Series: Journal of Physics: Conf. Series **1086** (2018) 012011 doi:10.1088/1742-6596/1086/1/012011

Spectrophotometer (Shimadzu, UV-2600 pc) at λmax = 664 nm. In each case, blank experiments were also conducted with the catalysts in the absence of light and without the catalysts when the solution containing the dissolved dye was illuminated. Repetitive photodegradation of MB during three consecutive cycles was performed with 1.0 g/L of catalyst at 10 mg/L dye concentration in order to investigate the stability of the photocatalysts. After each cycle, the catalyst was washed with distilled water and a fresh solution of MB was added before each photocatalytic run.

3. Results and discussion

3.1. Preparation mechanism of Sn/SnO$_2$

The possible formation mechanism of Sn/SnO$_2$ photocatalysts may include the following steps as: (1) conversion of A.C/SnCl$_4$ into A.C/Sn(OH)$_4$ by precipitation; (2) conversion of Sn(OH)$_4$ to SnO$_2$ by calcination at high temperature under open atmospheric conditions; (3) then, reduction of SnO$_2$ by activated carbon and CO to form SnO. As it is known that SnO is metastable, it will continue its reaction and passes into a stable state via reduction by activated carbon or CO to form metallic Sn and decompose into Sn and SnO$_2$ at high temperature; (4) oxidation of as-formed SnO and metallic Sn by O$_2$ in atmosphere.The simplified reaction processes may be summarized as below:

$$A.C/SnCl_4 + 4NH_4OH \xrightarrow{\Delta} A.C/Sn(OH)_4 + 4NH_4Cl$$

$$A.C/Sn(OH)_4 \xrightarrow{\Delta} A.C/SnO_2 + 2H_2O$$

$$SnO_2 + C \xrightarrow{\Delta} SnO + CO$$

$$SnO_2 + CO \xrightarrow{\Delta} SnO + CO_2$$

$$SnO + CO \xrightarrow{\Delta} Sn + CO_2$$

$$SnO + 2C \xrightarrow{\Delta} Sn + CO_2$$

$$2SnO \xrightarrow{\Delta} Sn + SnO_2$$

$$SnO + \tfrac{1}{2}O_2 \xrightarrow{\Delta} SnO_2$$

These results verified that activated carbon simultaneously play support and reducing agent role in formation of Sn/SnO$_2$ photocatalysts.

3.2. Evaluation of photocatalytic activity

To evidence the photocatalytic efficiency of the as-synthesized Sn/SnO$_2$ nanoparticles, experiments of the photocatalytic decomposition of methylene blue (MB) was performed as a test reaction according to the literature. Furthermore, blank experiments in the presence of irradiation without the photocatalyst were carried out to rationalize the photocatalytic activity of the as-synthesized Sn/SnO$_2$ photocatalyst. The degradation efficiency of the as-synthesized samples was defined as C/C_0, where C_0 and C are the initial concentration and the concentration of MB during the reaction time, respectively. Blank experiment results showed that MB could not be decomposed without the photocatalyst under UV irradiation (Figure 1a). With only catalyst, adsorption equilibrium study also showed that there was no appreciable adsorption of dyes on the catalyst.

In order to investigate the photocatalytic activity of Sn/SnO$_2$ photocatalyst, a set of parallel experiments were conducted with Sn/SnO$_2$ and SnO$_2$ photocatalysts. Figure 1(a) shows the photocatalytic activity of Sn/SnO$_2$ and SnO$_2$ photocatalysts for the degradation of MB under UV irradiation. It was found that the efficiency of SnO$_2$ was 41% after 210 min irradiation whereas for Sn-SnO$_2$ it was 96%. The composite nanoparticles showed higher photocatalytic activity than that of pure SnO$_2$.

For a better understanding of the photocatalytic efficiency of the as-synthesized samples, the kinetic analysis of MB degradation was discussed. It is well accepted that the photocatalytic decomposition of organic pollutants accords with a pseudo first-order rate equation of Langmuir–Hinshelwood which is

$$\ln (C/C_0) = -kt$$

where C_0 is the initial concentration of dyes and C is the concentration at reaction time. k represents the kinetic rate constant.

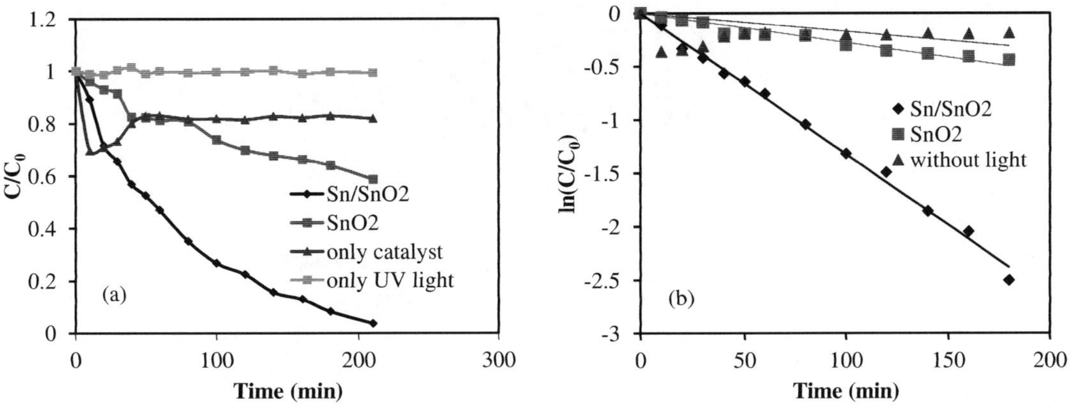

Figure 1. (a) Photocatalytic degradation of MB with as-prepared photocatalysts (b) kinetic study of methylene blue photodegradation

Degradation rate is quantitatively measured from the value of kinetic rate constant k. It can be understood how fast or slow a reaction is through the value of k. The rate constant, k was calculated by the linear regression of the slope of ln (C/C_0) vs. t plot. Fig. 2 shows the plots of ln(C/C_0) against t for different as-synthesized photocatalysts for the degradation of MB.The different values of k (0.0132 for Sn/SnO_2, and 0.0028 for SnO_2) shown in Figure 1(b) indicate the different degradability of the photocatalyst. The higher k value of Sn/SnO_2 photocatalyst indicates its higher photocatalytic degradation efficiency compared to bare SnO_2.

3.3. Photocatalytic Mechanism of heterostructure Sn/SnO_2 photocatalyst

Figure 2 shows the mechanism of photogenerated charge separation at the interface of heterojunction $Sn-SnO_2$ nanoparticles and its photocatalytic process. The higher photocatalytic activity of heterostructre Sn/SnO_2 nanoparticles is related to the role of Sn on the surface of SnO_2 nanoparticles. It is known that the work function of Sn (4.42 eV) is smaller than that of SnO_2 (4.84 eV). The Fermi energy level of Sn is higher than that of SnO_2 due its smaller work function. Accordingly, when the metallic Sn is attached to SnO_2 nanoparticles, the electrons will migrate from Sn to SnO_2 in order to achieve the Fermi level equilibration. As a consequence, the Fermi level of the semiconductor and metal in the interface is same that results in the formation of an electron depletion region and surface inward-bent band in the semiconductor. This electron depletion region formed at these Schottky barrier leads to an internal electric field at the interface of the heterojunction Sn/SnO_2 nanoparticles at thermal equilibrium. Under UV light irradiation, the electron in the VB of SnO_2 is excited to the CB by leaving a hole in the VB. The photogenerated electron hole pair can be separated effectively by the internal electric field developed at Sn/SnO_2 interface. The generated electrons tend to move to the positive side of the internal electric field; the electrons are transferred to the metal side take part in the reduction of dissolved oxygen forming superoxide radical anions, $O_2^{\bullet -}$, yielding hydroperoxy radicals $HO_2^{\bullet}$on protonation and finally $OH^{\bullet}$ radicals., while the photogenerated holes are transferred to SnO_2 particles and oxidize the surface hydroxyl group or physisorbed water molecules forming hydroxyl radical, $OH^{\bullet}$. $OH^{\bullet}$ radicals are a strong oxidizing agent well-known to decompose organic substrates as MB dye. Thus, an internal electric field caused the vectorial charge transport and transfer of electrons and holes into different directions and then retarded the photogenerated electron hole pair recombination, resulting in a better photocatalytic activity.

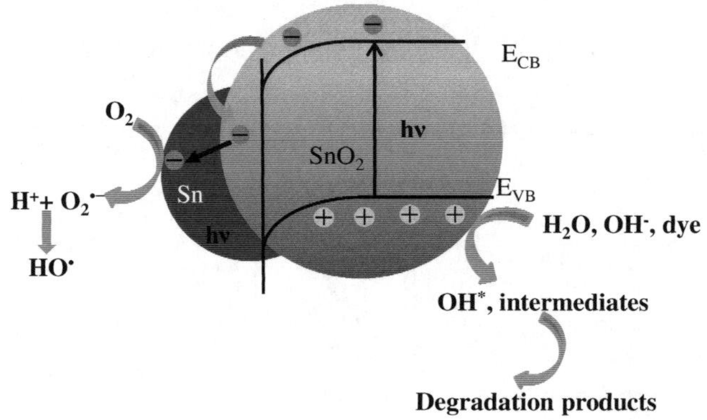

Figure 2. Schematic diagram of photogenerated charge separation by internal electric field at Sn/SnO_2 heterojunction and its photocatalytic process.

3.4. Effect of pH

The pH is an important parameter in the heterogeneous photocatalysis process. It influences the surface charge properties of the photocatalyst and therefore, the adsorption behavior of the pollutant. The pH of the dye solution was adjusted by adding NaOH or HCl. The role of pH on the photocatalytic degradation of MB was studied in three pH values (2.65, 6.50 and 10.50) at constant dye concentration (10 mg/ L) and photocatalyst amount 0.5 g/L.

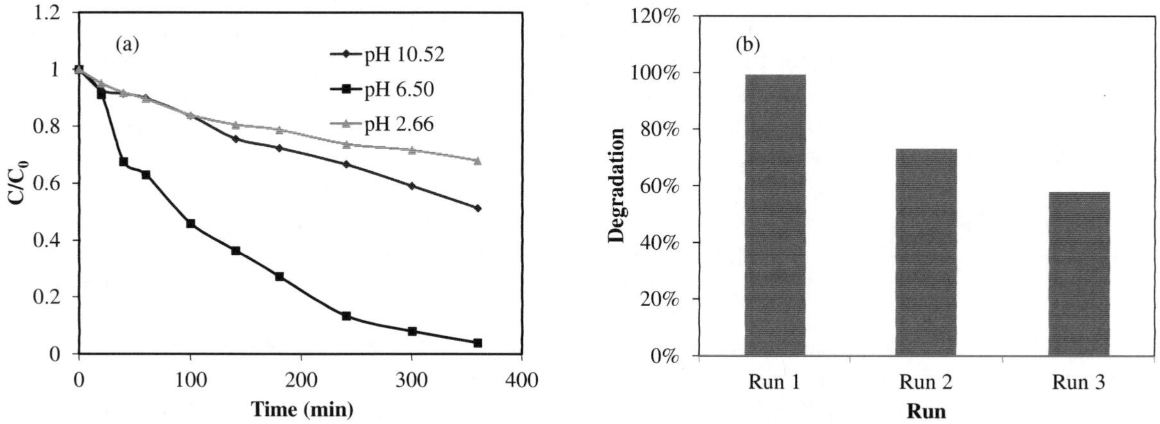

Figure 3.(a) Effect of pH of dye solution on the photocatalytic degradation of MB and (b) Cyclic runs in the photodegradation of MB using the Sn/SnO_2 photocatalyst under UV-light irradiation.

Figure 3(a) demonstrates the effect of pH on the photocatalytic degradation of methylene blue with different pH values. Figure 3(a) shows that MB photodegradation increased with increasing pH value from 2.65 to 6.50. Further increase in pH decreases the degradation efficiency. The degradation efficiency increases from 32% to 96% with increasing pH from 2.65 to 6.50 and subsequently the efficiency decreases to 49% with increasing pH to 10.50. The variation of solution pH changes the surface charge of the photocatalysts and shifts the potentials of catalytic reactions. As a result, the adsorption of dye on the surface is altered thereby causing a change in the reaction rate. Considering the positive charge of cationic MB dye in solution, the effect of pH on the photocatalytic degradation of MB can be rationalized on the basis of electrostatic adsorption model where cations are more readily accumulated at the negative sites on the heterostructure Sn/SnO_2. This can be explained on the basis of the point of zero charge (pHpzc) of Sn/SnO_2 particle. At lower pH, the surface of the catalyst is positively charged, but at higher pH it becomes negatively charged. Since MB is a cationic dye, high pH favors adsorption on the catalyst surface which results in the formation of thin film on the surface of the photocatalyst. This film acts as shield that inhibits the irradiated UV light to reach the catalyst surface resulting lower degradation of the dye at higher pH. On the other hand, at lower pH the electrostatic repulsion between the MB cations and positively charged oxide surface greatly reduces the adsorption of the MB dye resulting in a drastic decrease in degradation rate. The above discussion suggests that the neutral pH is favorable for the degradation of MB with the as synthesized Sn/SnO_2 photocatalyst.

3.5. Recyclability of synthesized Sn/SnO_2 for the degradation of methylene Blue

The reuse of catalyst is a critical criterion for commercialization of this technology, as this is a major factor contributing to the operating cost of the system. A photocatalyst should be resistive to photocorrosion, stable under different reaction conditions, and able to promote reactions efficiently upon repetitive use of Sn/SnO_2. The Sn/SnO_2nanocatalyst can be recycled and reused withoutsignificant loss of efficiency, which is a key requirement for a possible industrial application. Thus, the photocatalytic efficiency of the Sn/SnO_2 heterojunction photocatalyst for the 3-cycling reuse was 99.4, 73.2 and 58% after 360 min of reaction time, respectively (Figure 3 (b)). The degradation efficiency decreased because of the loss of catalyst during recovery for several runs.

4. Conclusions

Nanostructured Sn/SnO_2 photocatalysts have been successfully prepared by precipitation method followed by carbothermal reduction using activated carbon as reducing agent. The synthesized Sn/SnO_2 photocatalyats represented excellent photocatalytic activity for the degradation of MB under UV light irradiation. The pH of solution had a direct influence on the photocatalysis process and neutral pH that was favorable for the degradation of MB. Furthermore, the photocatalysts could be easily recycled without significant change in the catalytic activity which indicated the stability and reproducibility of the catalysts. This concept of semiconducting nanocatalysts with high photocatlytic activity should find industrial application in the future to remove undesirable organics from the environment.

Acknowledgments
The authors are thankful to the Ministry of Science and Technology, Bangladesh for financial support to conduct this research.

References

[1] The Bangladesh Responsible Sourcing Initiative: a new model for green growth 2014 The World Bank
[2] Dey S and Islam A 2015 *Resourc. Environ.* **5** 15-44
[3] Gaya U I and Abdullah A H 2008 *J. Photochem. Photobiol. C: Photochemis. Rev.* **9** 1-12
[4] Fujishima A and Honda K 1972 *Nature* **238** 37-38
[5] Chen X, Wu Z, Liu D and Gao Z 2017 *Nanoscale Res. Lett.* **12** 143
[6] Uddin, M T, Sultana Y and Islam M A 2016 *J. Sci. Res.* **8 (3)** 399-411
[7] Liu Y, Yu L, Hu Y, Guo CF, Zhang F M and Lou XW 2012 *Nanoscale* **4(1)**183-187
[8] Jing Z, Tan L, Li F, Wang F, Fu Y and Li Q 2013 *Indian J. Chem.* **52A** 57-62
[9] Linsebigler A L, Lu G and Yates J T 1995 *Chem. Rev.* **95** 735-758
[10] Zhang M, Sheng G, Fu J, An T, Wang X and Hu X 2005 *Mater. Lett.* **59** 3641-3644
[11] Rockenberger J, Felde U, Tischer M, Troger L, Haase M and Weller H 2000 *J. Chem. Phys.* **112** 4296-4304

IOP Conf. Series: Journal of Physics: Conf. Series **1086** (2018) 012012 doi:10.1088/1742-6596/1086/1/012012

Removal of methylene blue (MB) from waste water by adsorption on jackfruit leaf powder (JLP) in continuously stirred tank reactor.

T Ahmed[1], W Noor[1], O Faruk[1], M C Bhoumick[1] and M T Uddin[1]

Department of Chemical Engineering and Polymer Science, Shahjalal University of Science and Technology, Sylhe-3114, Bangladesh.

E-mail: mcbhoumick.sust@gmail.com

Abstract. Dye contaminated waste water possess serious threat to the goal of reaching sustainable environment. Bio adsorbents can be utilized very effectively in removing dye ingredient from wastewater due to their high kinetics and capacity. The study was focused on observing bio adsorbent (jackfruit leaf powder) activity for Methylene Blue adsorption in a Continuous Stirred Tank Reactor (CSTR) for wastewater treatment.Batch and kinetic study of adsorbent were performed. The maximum adsorption capacity at equilibrium was found to be 271 mg/g and adsorption followed Langmuir isotherm. The kinetics study provided evidence of instantaneous adsorption while the zero point charge of the adsorbent was 3.9. The FTIR of the JLP-MB system was carried out to explain the adsorption phenomena with representing groups. The CSTR study was performed under various parameters such as, effect of flow rate, adsorbent dosage, CSTR arrangement and initial concentration. The dye stream first treated in 1st reactor and then its effluent passes into the 2nd reactor and treated again. It was found that the single CSTR removes 30-45 % of the dye depending on the initial concentration, while the series effect removes 70-85 % of the dye overall.

1. Introduction:

Water is the foundation of life on planet and one of the most important constituents of modern industrialization. Considering both volumes discharged and effluent composition, the wastewater generated by the textile industry was rated as the most polluting among all industrial sectors [1, 12]. Many types of dye are used in textile industries such as direct, reactive, acid and basic dyes. Most of these dyes represent acute problems to the ecological system as they considered toxic and had carcinogenic properties, which make the water inhibitory toAquatic life [6]. The effluents from dye manufacturing and consuming industries are highly colored coupled with high chemical and biochemical oxygen demands (COD and BOD) [2]. Discharge of such effluents imparts color to receiving streams and affects its aesthetic value.Color hinders thepenetration of sunlight into waters, retards photosynthesis, inhibits the growth of aquatic biota and interferes with gas solubility in water bodies [9]. Dyes may also be problematic if they are broken down anaerobically in the sediment, as toxic amines are often produced due to incomplete degradation by bacteria [7]. Dyes laden wastewater is usually treated by physical or chemical processes. All of these methods have different color removal capabilities, capital costs and operating rates. Among these processes, adsorption has been found to be superior to other techniques for wastewater treatment in terms of initial cost, simplicity of design, ease of operation and insensitivity to toxic substances [9]. A small amount of adsorbent can adsorb adequate amount of adsorbate. Commercial activated carbon is an excellent adsorbent in terms of adsorption capacity but due to its high cost, adsorbents derived from cellulosic sources are extensively studied. Researchers have investigated dye removal from aqueous solutions using natural materials or agricultural wastes such as palm shell [17], coconut coir[18] sawdust[19], jute fiber[20], rice husk[21]mango leaf powder [22]etc. as low cost biosorbents[16]. In this present study Jackfruit leaf powder (JLP) was used as an adsorbent for MB adsorption. The present study was mainly focused on how to treat dye containing waste water by adsorption via bio adsorbent in Continuous Stirred Tank Reactor (CSTR). CSTR was advantageous over batch or column system in removal of waste constituents from wastewater stream because of continues steady state operation. The objective of this study was to find an efficient method of dye removal from waste water in continuous basis by series CSTR system for MB-JLP adsorption as a model.

2. Experimental Methods:

2.1 Materials:

Jackfruit leaf powder was used to study adsorption of methylene blue in CSTR. Leaves were collected from local area. After collecting the leaves, they were thoroughly cleaned to remove dirt and other earthy particles. Then the leaves were dried in oven at 105°C overnight. Dried leaves were crushed and boiled in water until the removal of its mostlignin and coloring components. Boiled leaves then filtered and dried in oven at 105°C for 24 hours and sieved to particles size 45-500μm and stored in a plastic jar for using as an adsorbent.Methylene Blue (Fig-1) was purchased from Merck and used as received without further purification. The stock solution of MB was prepared with a concentration of 1000 ppm (1000mg/L) and all other working solutions were prepared by diluting the stock solution with distilled water.

2.2. Methods:

To study the effect of important parameters e.g. adsorbent dosage, initial dye concentration, and contact time [1, 3] on the adsorptive removal of MB, batch experiment was conducted on 25±2°C. In this study six 250 ml stopper bottles were used. Each of them was filled with 200 ml of methylene blue solution of seven different concentrations. A fixed amount of adsorbent was added to each solution. The samples were then shaken with a flask shaker (Stuart Scientific Co. Ltd, Model: SF1, UK) at 450osc/min for 6hrs. Within this time the system reached equilibrium. The solutions were then centrifuged for 5-10 min to settle down the adsorbent. The dye solution was then separated and the concentration of the solution was determined with UV-vispectrophotometer (UV-2600 PC, Manufacture: shimadzu, Made in Japan) at 664 nm wavelength.
The adsorption density q_e can be expressed as

$$q_e = \frac{V(C_0 - C_e)}{W} \quad \text{.........} 1$$

Where, C_0 and C_e are the liquid phase concentration (mg/L) of MB at initial time and any time t respectively. V was the volume of the solution (L) and W was the mass of the adsorbent (gm).For adsorption isotherm, where the equilibrium density q_e (mg/gm) was calculated by Eq (1).Theoretical equation for calculating solute (adsorbent) concentration on the CSTR and outlet with time was;

$$(C_{in} - C) = (C_{in} - C_0)\, e^{-\frac{t}{\tau}} \quad \text{.........} 2$$

If no solute was fed to the CSTR with inlet stream then $C_{in} = 0$ and equation 2 reduces to,

$$C = C_0 e^{-\frac{t}{\tau}} \quad \text{.........} 3$$

During single CSTR operation; initially the CSTR was filled with a dye solution of a fixed concentration up to the outlet pipe where total solution volume was 2L. Fixed dose of adsorbent W (gm/L) was added to the CSTR at time t=0. Dye solution from the storage tank was added to the CSTR at a constant flow rate. The flow rate was maintained with the help of a valve. The adsorbent concentration in the CSTR was kept constant by adding the amount lost every 5 minutes. This amount was calculated from Eq. (3). The overflow solution was collected every 5 min and centrifuged to settle the adsorbent. The sample was then taken into test tube and the concentration was measured with UV-vis spectrophotometer. Effluent was passed through the discharge of 1st CSTR.When two CSTR are in series operation; steps are the same as single CSTR operation. But rather passing all effluent from 1st CSTR through discharge, the stream was directed into 2nd CSTR by closing ball valve of 1st CSTR discharge. When 2nd CSTR was filled with effluent of 1streactor, same previous steps were maintained as single CSTR for 2nd reactor too.

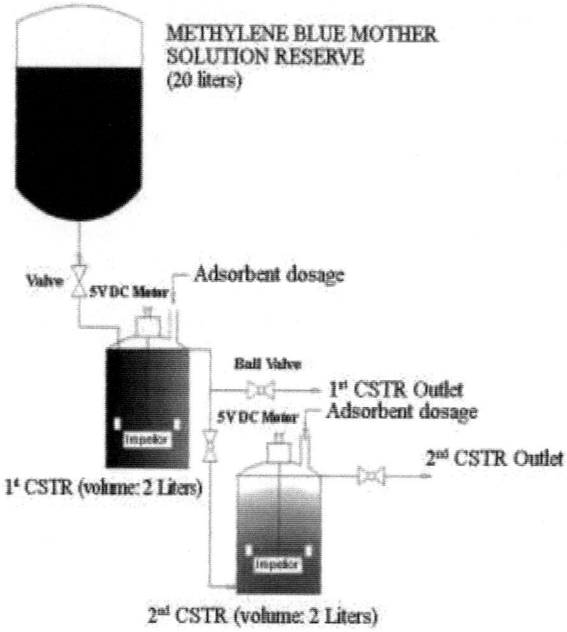

Figure 1. Complete diagram of series CSTR system for adsorption of MB onto JLP

3. Results and Discussion

3.1. Particle size distribution

Particle size distribution was measured by sieve shaker (Vibratory Sieve Shaker, Analysette) of five sieve size tray (500, 250, 125, 63, 45 μm). 30gm adsorbent was put in the shaker and run for 3 hours. And the remaining on the top of sieve tray was weighted carefully. The distributions of particle size are shown on figure 2.

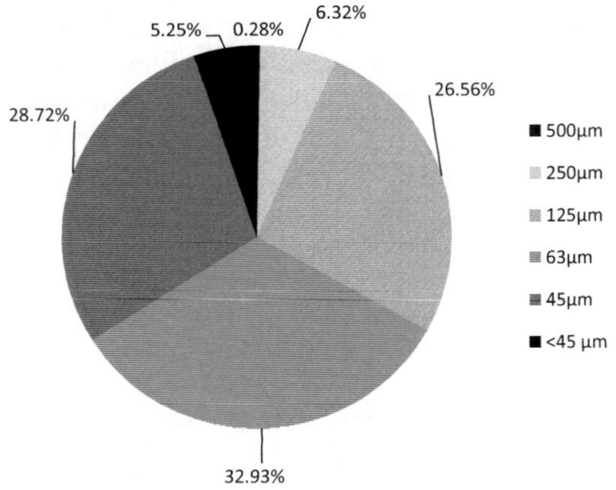

Figure 2. Particle size distribution

3.2. Zero Point Charge

The surface charge, Q was calculated from the experimental titration data according to the following equation

$$Q = 1/w\ (C_A\text{-}C_B\text{-}[H^+]\text{-}[OH \qquad \text{.......}4$$

where, w- dry weight of adsorbent in aqueous system (g/L); C_A is the concentration of added acid in aqueous system (mol/L); C_B is the concentration of added base in aqueous system (mol/L); $[H^+]$is the concentration of H^+(mol/L); $[OH^-]$ is the concentration of OH^- (mol/L). The pH value at the point of zero charge was then determined by plotting Q versus pH.The plot on figure 3 illustrates the surface charge property of JLP adsorbent particles. It was found that at pH 3.9 the net charge on the JLP surface was zero. Above 3.9 the surface charge is negative and below vice versa. MB is a positive type of dye, hence higher pH favors adsorption.

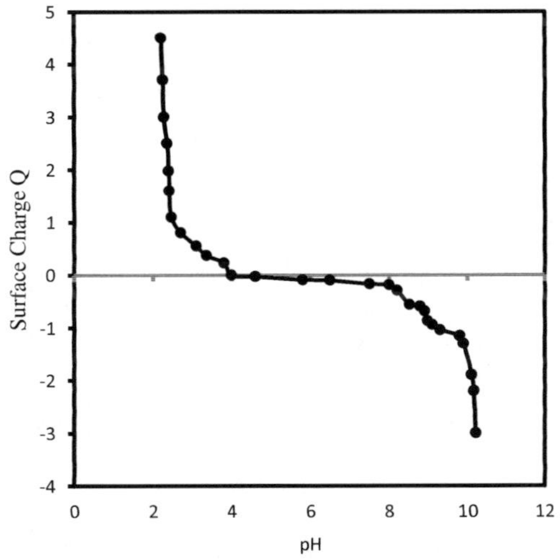

Figure 3. Zero Point Charge of JLP adsorbent.

3.3. Fourier Transform Infrared Spectroscopy of adsorbent

The FTIR spectrums of JLP before adsorption of MB are shown in Figure 4 and that of JLP after adsorption are revealed in Figure 5. As shown in Figure 4 the spectra display a number of absorption peaks, indicating the multifaceted nature of the material examined. The FTIR spectroscopic analysis dictated broad bands at3418 cm^{-1}, representing bonded OH groups[11]. The band observed at 1743 cm^{-1} was assigned to a carbonyl band (CO) of pectin ester[15], while the peak at 1616 cm^{-1} was attributed to C=O stretching of carboxylic acid with intermolecular hydrogen bond. The peak observed at 1520 cm-1 corresponds to secondary amine group[16].FTIR analysis indicates that JLP was represented by functional group OH, COOH, NH, CO, that could be potential adsorption site for MB.

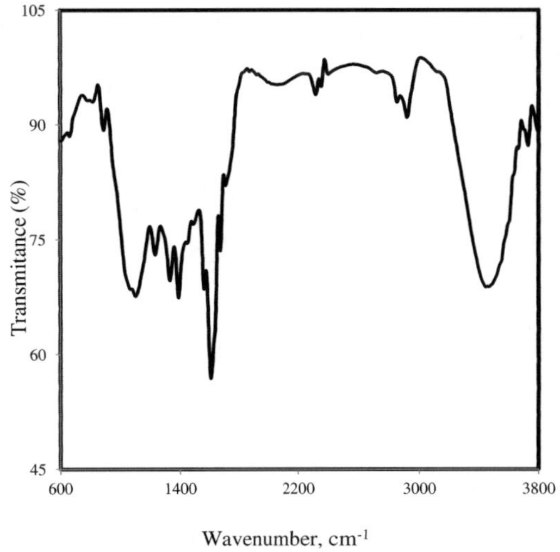

Figure 4. FTIR of JLP before adsorption

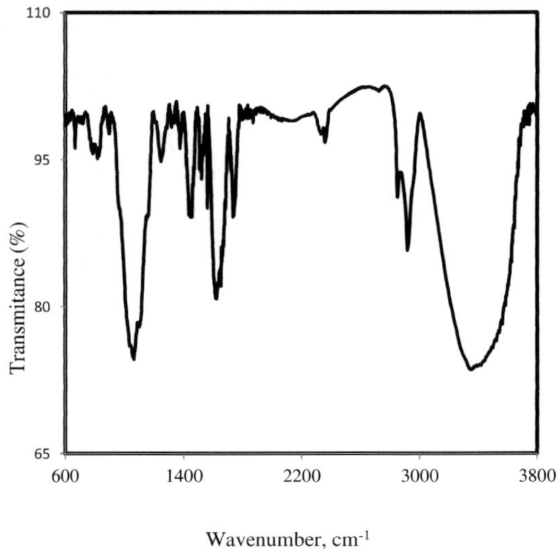

Figure 5. FTIR of JLP after adsorption

3.4. Adsorbent Capacity

In batch study, it was seen that with increasing concentration of MB solution, the capacity of adsorbent tends to increase. In the figures 3and 4, it shows that capacity increases with growing concentration; capacity increases for a certain amount adsorbent dosage.

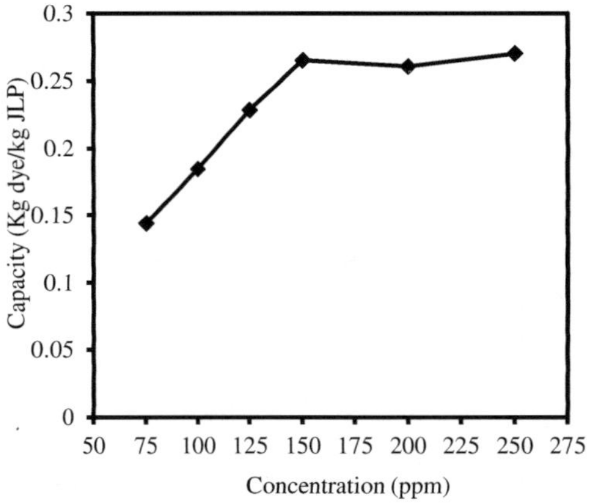

Figure 6. Adsorbent capacity vs. initial dye concentration (Volume of solution = 200ml, adsorbent added = 0.1gm, initial concentration 75, 100, 125, 150, 200, 250ppm)

3.5. Langmuir Adsorption isotherm:

The Langmuir adsorption isotherm was valid for monolayer adsorption onto a surface containing a finite number of ideal sites. This model assumes uniform energies of adsorption onto the surface and no transmigration of adsorbate in the plane of the surface. The Langmuir adsorption isotherm for solid-liquid adsorption system was represented by the following equation

$$\frac{1}{q_e} = \frac{1}{q_\infty K} \cdot \frac{1}{C_e} + \frac{1}{q_\infty} \qquad \text{.......5}$$

Where, q_∞ and K was the adsorption capacity (Kg adsorbate/Kg adsorbent) and rate of adsorption respectively. A plot of the $1/q_e$ vs. $1/C_e$ was for MB adsorption on JLP was presented on Fig 6. The Langmuir constant are calculated by equation 5 and tabulated in Table-1. Form Figure 7, it can be seen that adsorption of MB onto JLP was best fitted with Langmuir isotherm. The maximum sorption of JLP for MB was found 271mg/gm. Table 2 shows a comparison of other low cost adsorbent for uptake of MB.

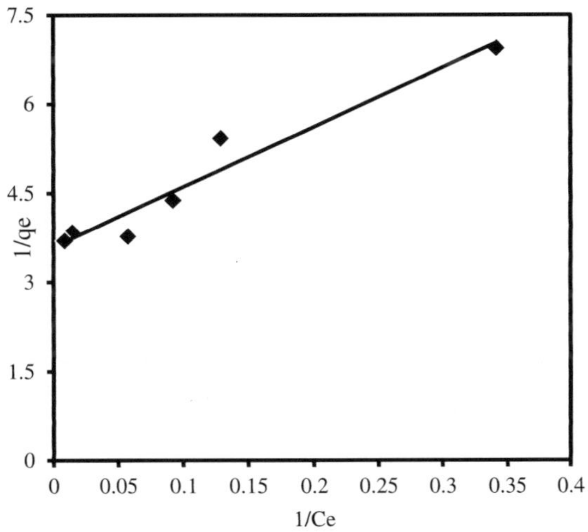

Figure 7. Langmuir isotherm for methylene blue adsorption onto JLP (Volume of solution = 200ml, adsorbent added = 0.1gm, initial concentration 75, 100, 125, 150ppm, Temperature 24±2°C)

Table 1. Langmuir isotherm constants

System	q_∞ (kg/kg)	K (m^3/kg)	R^2
MB-JLP	0.271	0.3307	0.95

IOP Conf. Series: Journal of Physics: Conf. Series **1086** (2018) 012012 doi:10.1088/1742-6596/1086/1/012012

Table 2. Adsorption capacity of different adsorbents found in literature

Adsorbent	Capacity (mg/gm.)	Authors
Bio-sludge ash	1.599	(Chih-Huang Weng, 2005)
Activated carbon of rattan sawdust	294.1	(B.H. Hameed A. A., 2006)
Cedar sawdust	142.36	(Hamdaoui, 2006)
Crushed brick	96.61	(Hamdaoui, 2006)
Bituminous coal-based activated carbon	588	(Emad N. El Qada, 2006)
Peanut hull	68.03	(Renmin Gong, 2005)
Clay	58.2	(A. Gürses, 2003)
Jackfruit leaf powder	271	Present Study

3.6. Kinetics Study

Effects of contact time on adsorption of dye were studied for the different initial concentration of dye with fixed Adsorbent dosage Figure 8 represents methylene blue JLP system. It was seen from the figure that the initial rate of adsorption was high for both the system and 70-85% of dye was removed within first 10 min of adsorption. It represents adsorbent's high kinetic ability. After that, the rate was decreasing gradually leading to equilibrium. This was due to the high capacity of the adsorbent and the availability of dye molecules at the adsorbent interface. When the surface active sites of adsorbent are covered fully, the extent of adsorption reaches to a limit resulting in saturated adsorption.

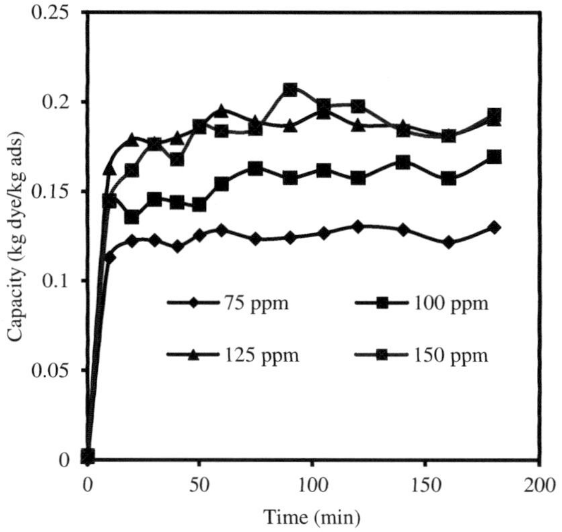

Figure 8. Kinetics study of Methylene blue adsorption onto JLP. (Volume of solution = 200ml, adsorbent added = 0.1gm, initial concentration 75, 100, 125, 150ppm, Temperature 25±2°C; shaking speed = 350 osc. /min)

IOP Conf. Series: Journal of Physics: Conf. Series **1086** (2018) 012012 doi:10.1088/1742-6596/1086/1/012012

4. CSTR performance study

4.1. Single CSTR Study

For this study, a fixed amount of adsorbent (1gm.) was used. Adsorbent makeup dosage varied with flow rate. Higher the flow rate, higher the makeup dosage. Adsorbent make up dosage was calculated by equation 3. For single CSTR, performance was studied for three different flow rates and every time, the reduced concentration of effluent dye was in the desired range with satisfactory dye removal. In this study, initial adsorbent dosage was maintained all over the reactor operation. And make up dosage was applied in every five minutes after the sample withdrawal.

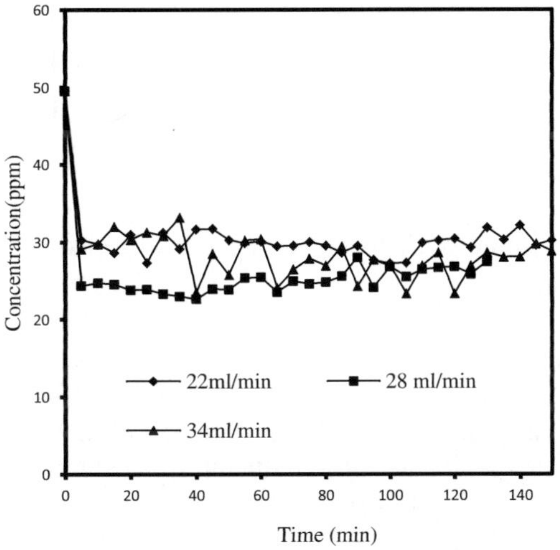

Figure 9. Effluent concentration curve of Single CSTR operation for adsorption of MB onto JLP.

4.2. Series CSTR Study

Two equal volume CSTR was connected in series to see the overall removal of MB. Effluent from 1st CSTR enters into the inlet of 2nd CSTR. Same parameters (flow rate, adsorbent make up dosage) were cautiously maintained in both reactors to get the performances equally. It was seen that, series effect increases the removal of methylene blue quite significantly. While the first reactor removed 40-50 % of the initial dye, with the addition of second one the removal percentage ascended to around 75-90%. Figure 10 summarizes the whole study.

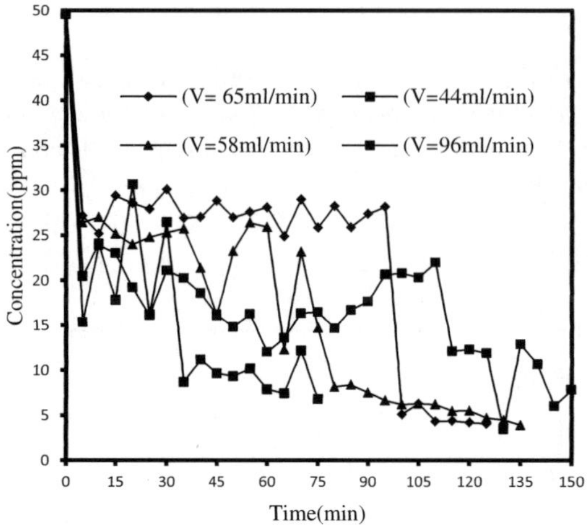

Figure 10. Series CSTR operation for fixed amount adsorbent for adsorption of MB onto JLP.

5. Conclusions

This study was primarily focused on finding an alternative potential system to treat wastewater by adsorption with high rate and efficiently in CSTR. Due to continuous steady state mode of operation of CSTR; for adsorption purpose, it was required to use an adsorbent with high kinetics and sorption capacity. Jackfruit leaf powder (JLP), for its high kinetics, and comparatively high capacity makes it an outstanding choice for using in CSTR system for treating dye from the waste stream. It was found that, in CSTR, adsorption can be possible with a considerable ease by using low-cost natural adsorbent. According to Langmuir isotherm, MB adsorption on JLP was favorable. The maximum sorption capacity of MB onto JLP has been found 271mg/gm. it was relatively higher than other natural low-cost adsorbent except for activated carbon. Other adsorbent with higher kinetics and capacity can be used in CSTR to treat dye containing waste stream by adsorption method easily and proficiently.

References

[1] Gürses A, Karaca S, Doğar C, Bayrak R, Açikyildiz M andYalçin M. 2004*J Colloid Interface Sci.***269** 310
[2] Hameed B H, AhmadA L, Latiff K N A2006*Dyes and Pigm.***75** 143
[3] Hameed B H, Din A T M and Ahmad A L 2006*J. Hazard. Mater.***141** 819
[4] Considine D M andConsidine G D 1995 Van Nostrand's Scientific Encyclopedia *Springer*
[5] Weng C -H, Pan Y –F 2005*Colloids and Surfaces A: Physicochem. Eng. Aspects* **274**154–162.
[6] Dai M. 1994*J. ColloidInterface Sci.* **164** 223–228.
[7] E. Weber, N. W. (1987). Kinetics studies of reduction of aromatic azo compounds in anerobic sediment/water systems. Environ. Toxicol.Chem. 6.
[8] Qada E N E, Allen S J and Walker G M2006*Chem. Eng. J.* 124 103
[9] Hamdaoui O 2006J. Hazard. Mater.**135**264–273.
[10] Avom J, Mbadcam J K, Noubactep C and Germain P 1997Carbon **35** 365
[11] Uddin M T,Rukanuzzaman M,Khan M R and Islam M A 2009Indian *J. Chem. Technol.***16**142
[12] Vandevivere P C, Bianchi R, nadVerstraete W1998*J. Chem. T echnol. Biotechnol***72**289-302.
[13] Gonga R, Li M, Yang C, Sun Y and Chen J 2005*J. Hazard. Mater.***121** 247
[14] Bukallah S B, Rauf M A and Ali S S2006 *DyesPigm.* **74** 85
[15] Uddin M T 2009 Jackfruit Artp. *Indian chemical* 8

IOP Conf. Series: Journal of Physics: Conf. Series **1086** (2018) 012013 doi:10.1088/1742-6596/1086/1/012013

Thermal performance comparison of flat plate evaporator loop heat pipe operating between horizontal condition and gravity assisted condition

A R Tuhin[1], P H Huynh[2], K Z Htoo[2], K Kariya[3] and A Miyara[3]

[1] Exchange student, Graduate School of Science and Engineering, Saga University, 1 Honjo-machi, Saga,840-8502, Japan and Undergraduate Student, Department of Mechanical Engineering, Chittagong University of Engineering & Technology, Bangladesh.
[2] Graduate School of Science and Engineering, Saga University, 1 Honjo-machi, Saga,840-8502, Japan.
[3] Department of Mechanical Engineering, Saga University, 1 Honjo-machi, Saga, 840-8502, Japan.

Email: metuhin12@gmail.com*, hphien@hcmut.edu.vn, kyawzinhtoo1992@gmail.com, kariya@me.saga-u.ac.jp and miyara@me.saga-u.ac.jp

Abstract. With huge developments of science and engineering especially information and communication technology, loop heat pipe (LHP), a passive two-phase cooling device, can be considered one of the promising heat transfer devices with effective thermal conductivity. In this study, a new evaporator structure is fabricated and experiments are carried out horizontal and gravity assisted operating condition. Water is selected as working fluid while sintered stainless steel wick is installed as capillary structure of LHP. The Results show that, gravity assisted LHP has higher performance compared to horizontal LHP and it can keep stable operating state at heat load 520 W without dry out phenomenon.

1. Introduction

Nowadays, conventional heat pipe (HP) is used widely and successfully in cooling electronic equipment. With their miniature and multinational tendency, heat flux generated from electronics increases day by day. For example, there is a survey that 50-1000 elements have been installed per chip in the 1960s but the prediction said that it will be about 20 billion per 1 cm^2 in 2020 (1). Therefore, it is necessary to study on new cooling methodologies with new challenges of heat transport capacity. Although both HP and LHP is a heat transfer device which operation principle based upon phase changing process and utilizes the capillary or gravity force to complete the circulation of working fluid, there is a difference between normal HP and LHP. To compare with other HP, the LHP has the higher capacity of heat transfer and flexible characteristics because of the absence of the wick structure in the connection lines as well as separation of vapor and liquid lines. It can eliminate the occurrence of entrainment or flooding limitation that restricts condensed liquid from returning evaporator. Moreover, the LHP comprises a twophase working fluid reservoir which is known as compensation chamber. This compensation chamber plays an important role to determine the pressure and temperature of LHP. In addition, fine porous sintered wick is often selected to create

strong capillary force for operating under horizontal condition as well as favourable gravity assisted condition. Until now, LHP with cylindrical evaporator has been utilized successfully in cooling electronics of extra terrestrial applications. To compare with flat rectangular evaporator LHP, the main disadvantage of cylindrical evaporator LHP is having higher thermal resistance between evaporator and heat source. To reduce the thermal contact resistance, even if thermal resistance sheet is inserted between evaporator and heater source, it will little decrease the thermal resistance, but at the same time it will increase the weight of electronics device. Although both flat rectangular evaporator LHP and disk type evaporator LHP have the almost same performance and can reduce the thermal contact resistance, but most of the electronics devicethat require cooling have a flat surface. In summary, flat rectangular evaporator LHP can use the proper area of electronics device as well as reduce the thermal contact resistance. Therefore, with developing trend of electronics equipment especially CPU, data centre etc…LHP with flat rectangular evaporator are now again considered as potential solution.
From the previous points, there are many studies are going on to find an efficient, low cost and long term reliable electronics cooling solution with minimum maintenance demand. Zhou et al. (2) designed a mini LHP for cooling mobile electronics, which consists of a heating power of 11 W and an evaporator temperature of 97.2 0C. Ayaka Suzuki et al. (3) experimented on cooling performance of a JEST (Jet Explosion Stream Technology) type LHP. They confirmed that maximum heat transport rate was 780 W, when heat flux was 195 W/cm2 and obtained a total thermal contact resistance of 0.13 K/W. Cheiko Kondou et al. (4) carried out experiment on looped thermosyphon performance with volatile fluids such as R1234ze(Z) and R1234ze(E) as working fluids. In their study, they conclude that by using super-hydrophilic boiling surface, nuclear boiling heat transfer can be enhanced. Shioga and Mizuno (5) investigated a MLHP with water as working fluid for thermal management of smart phones.

In this study, a flat-rectangular evaporator copper LHP is fabricated and the experiment was conducted to measure the performance under horizontal condition which is also known as 'no gravity' condition and gravity assisted condition. Sintered stainless steel is used and water is selected as working fluid as water can operate in low pressure and has good compatibility with copper. As shown in Figure 1, a system of fins or crossing grooves is machined to confirm the heat transfer from heater to the wick and create the paths for vapor flow out sufficiently.

Figure 1. Evaporator with sintered stainless-steel wick

2. Experiment setup and condition

2.1. Experimental setup

Figure 2 demonstrates the schematic diagram of the experiment. Both horizontal and gravity assisted LHP are consisted as a loop that connected evaporator, vapor line, condenser and liquid line, respectively together. Evaporator, vapor line, condenser and liquid lines are placed in same elevation in case of horizontal LHP. On the other hand, when experiment was carried out in gravity assisted condition, vapor and liquid line was set up vertically and condenser was set up horizontally above evaporator and connected with vapor line and liquid line. Other experimental apparatus kept same orientation in both cases. Four cartridge heaters are inserted inside the copper heating block to heat the evaporator section. To reduce the thermal contact resistance, a thin layer of thermal conductivity grease is filled between the upper surface of heating block and the lower surface of evaporator. AC source is converted into DC source by the YAMABISHI MVS – 520 Volt slider and heat load is adjusted while magnitude is monitored on the YOKOGAWA WT230 digital power meter. On the other hand, condenser is cooled by water which circulate between condenser and ADVANTEC LV – 400 constant temperature circulator. As shown in figure 3, three (0.5mm diameter) K type thermocouples are inserted into the heating block to calculate the heat flux and temperatures (T1, T2, T3) at the top surface of heating block Ts1. One (1mm diameter) K type thermocouple has been inserted to estimate the temperature (T4) at bottom surface of evaporator. Moreover, temperatures at evaporator outlet (Teo), condenser outlet (Tco) as well as compensation chamber inlet (Tcci) are determined by three K type thermocouples and inserted directly on the path of working fluid. The mass flow rate and the temperature difference of cooling water when flowing through the condenser are measured by the MASSMAX MMM7150K mass flow meters and two K type thermocouples Twa-i, Twa-o respectively. Another thermocouple Ta measures the surrounding air temperature. Except for Ta, thermocouples are calibrated to ensure that error is smaller than ±0.05°C. All measuring data is collected automatically by using Keithley 2071 Data Acquisition System which interfaces with PC by Exelink software. Table 1 shows the LHP's parameters in detail.

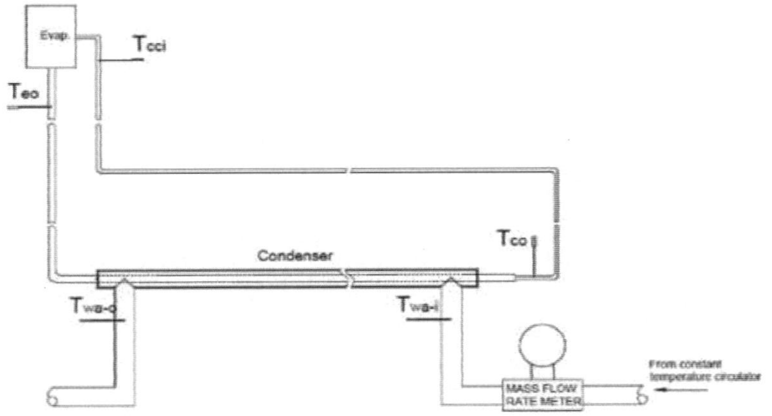

Figure 2. Schematic diagram of experiment

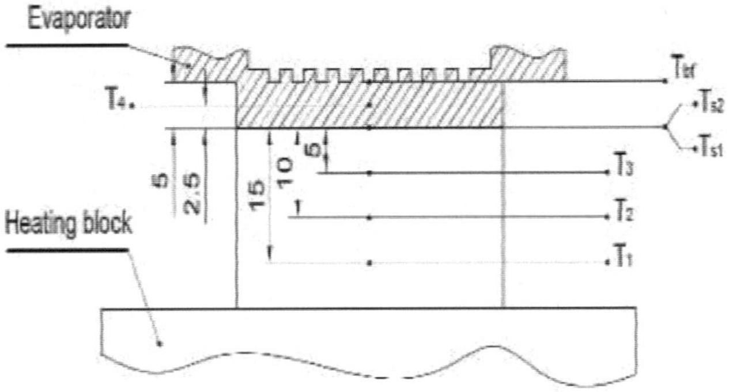

Figure 3. Positions of thermocouples inside heating block and base body of evaporator

IOP Conf. Series: Journal of Physics: Conf. Series **1086** (2018) 012013 doi:10.1088/1742-6596/1086/1/012013

Table 1. Main parameters of LHP in experiment

Evaporator	
Material	Copper
Length, mm	80
Width, mm	70
Height, mm	25
Active area, mm^2	60 x 45
Fin geometry	
Cross area, mm^2	2 x 2
Height, mm	1.5
Fin pitch, mm	4
Wick Structure	
Material	Stainless steel
Pore radius, μm	63
Porosity, %	36 – 48
Bulk volume, mm^3	50 x 41 x 5
Vapor line	
Length, mm	700
OD/ID, mm	6.35/4.35
Condenser	
Length, mm	600
OD/ID, mm	6.35/4.35
Liquid line	
Length, mm	1300
OD/ID, mm	3.2/1.7
Working fluid	Water

2.2. Experimental condition

The insulation of the heating block, evaporator, vapor line and condenser section are done carefully to eliminate the heat loss from heater as well as influence of environment on the result. During experiment, room temperature is kept constant around 25oC. The constant temperature circulator maintains the inlet temperature Twa-i and mass flow rate of cooling water are of 28.5oC, 30 kg/h respectively. The heat power input is increased until the temperature (T4) reaches 100oC. Besides, to guarantee vacuum condition as well as proper amount of working fluid charged in LHP, the charging system consisting liquid tank, glass level indicator and stop valve as shown in Figure 4 was established, and the charging procedure was conducted carefully as following steps.

IOP Conf. Series: Journal of Physics: Conf. Series **1086** (2018) 012013 doi:10.1088/1742-6596/1086/1/012013

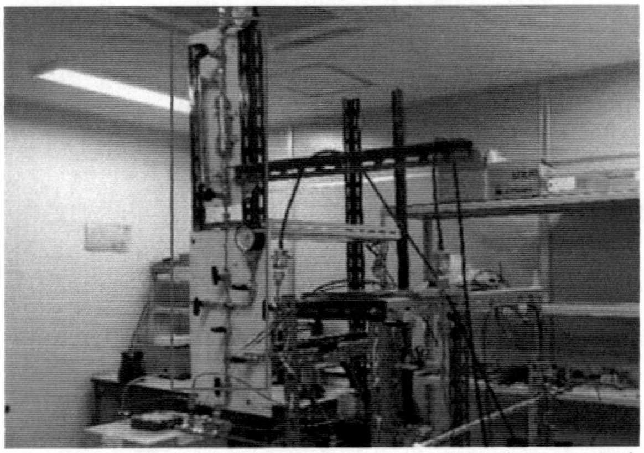

Figure 4. Connection between charging system (red rectangular) and LHP

Firstly, both of LHP and charging system was vacuumed to remove all non-condensing gases. Then, water was charged into charging system while LHP was disconnected by the stop valve. During this process, vacuum condition of charging system was lost; hence, it is necessary to vacuum the charging system again. Finally, water is charged into LHP by opening the stop valve (connection between charging system and the LHP). The amount of charging water was determined from glass level indicator. Moreover, before charged into charging system, the purified water was also boiled to reduce the concentration of dissolve gases inside its volume.

2.3. Data reduction

Heat flux and the heat flow rate flowing through the heating surface to active area of evaporator

$$q = k\frac{T_1 - T_2}{\delta} = k\frac{T_2 - T_1}{\delta} = k\frac{T_1 - T_3}{2\delta} \quad (1)$$

$$Q = qA \qquad (2)$$

Heater surface temperature T_{s1}

$$T_{s1} = T_1 - 3\frac{q\delta}{k} \quad (3)$$

This temperature is considered as electronics temperature. With different electronic devices, the limits of this temperature are different. From (6), this value is often recommended to be at 85oC for reliable and effective operation of electronics such as microprocessors, processers etc.

Total thermal resistance Rt

$$R_t = \frac{T_{s1} - T_{wa-i}}{Q} \quad (4)$$

3. Results and Discussion

3.1. Heat transfer capacity

Experiments showed that operating condition has the significant effects on varying heat transfer capacity of LHP. Under horizontal condition, fluid circulation in an LHP only driven by the capillary force where gravity assisted condition, fluid circulation in an LHP driven by both capillary and gravitational forces. With the condition, temperature (T_4) at the base body of the evaporator, is lower than 100 ^{0}C, the gravity assisted LHP can work until the heat load reaches 520 W. On the other hand, horizontal LHP can transfer only 50 W from heater to the heat sink. For the horizontal LHP, the capillary condition can describe in Eq (5). Vapor phase pressure drop such as vapor grooves, vapor line, and condenser line pressure drop is denoted by Δp_v. On the other hand, pressure drop when liquid flows through liquid line and wick structure is denoted by Δp_l. The length of the connection lines contributes to the increment of total pressure drop, causes the capillary limitation happens and this explains why horizontal LHP reduces heat transfer capacity.

$$\Delta p_c = \frac{2\sigma}{r} > \Delta p_v + \Delta p_l (5)$$

3.2. Ts1 and temperatures at different positions inside LHP change with heat load Q

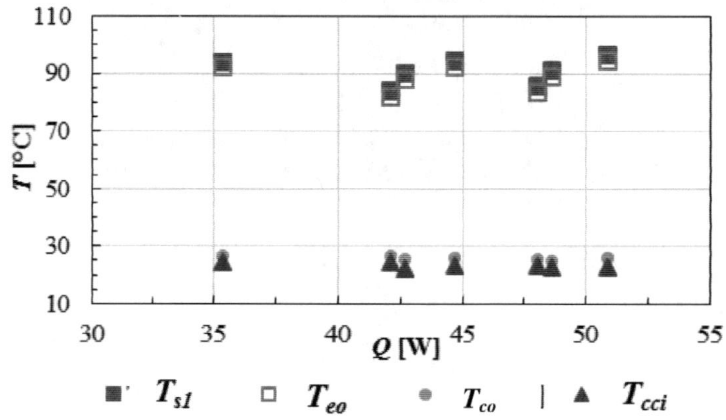

Figure 5. Temperatures vs heat load in the case of horizontal LHP

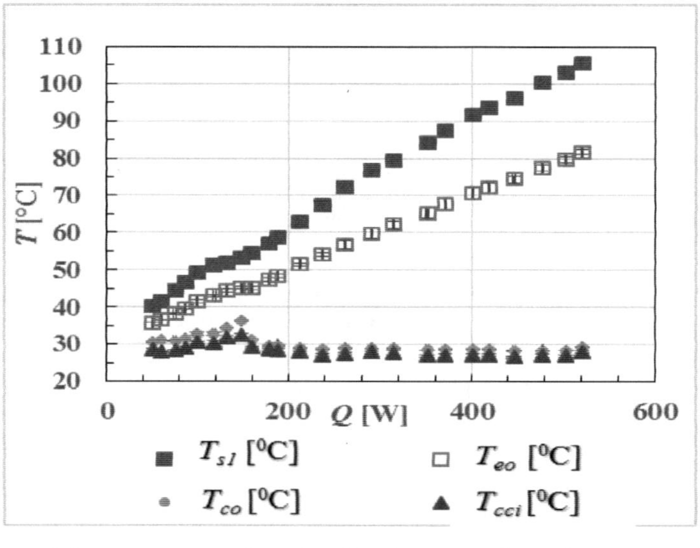

Figure 6. Temperature vs heat load in the case of gravity assisted LHP

The values of T¬s1, Teo, Tco and Tcci of the horizontal and gravity assisted LHPs are demonstrated on figure 5 and figure 6 when operating at different heat load conditions. Both cases, Tco is a little lower than T¬s1 which confirms that circulation happens inside the LHP. Moreover, Tco and Tcci are nearly equal together. This result shows that there is no reverse flow from compensation chamber to condenser that will cause the sudden increment of Tcci. In horizontal LHP, T¬s1 is almost higher than 85 0C although heat load is lower than 55 W. Besides, changing tendency of this temperature is not regular with heat load. However, gravity assisted LHP can keep T¬s1 lower than 85 0C until heat load reaches 350 W.

3.3. Total thermal resistance vs heat load

To understand more clearly performance of both horizontal and gravity assisted LHPs, the values of total thermal resistance Rt according to heat load are demonstrated in Figure 7. In horizontal condition, Rt is always higher than 1.2 K/W and it has the reducing trend with heat load range from 35 W to 55 W. On the other hand, gravity assisted condition, the minimum value of Rt is 0.133 K/W and the operation characteristics can be divided into two types that are variable thermal resistance and constant thermal resistance. From this, it is observed that in case of gravity assisted condition (when working at high heat load), the relation between heater surface temperature (T¬s1) and heat load (Q) is nearly linear. The higher values of Rt under low heat load conditions can be explained on the basis of upon the small evaporating rate; as a result, it is difficult for evaporating vapor flows out vapor groove to condenser.

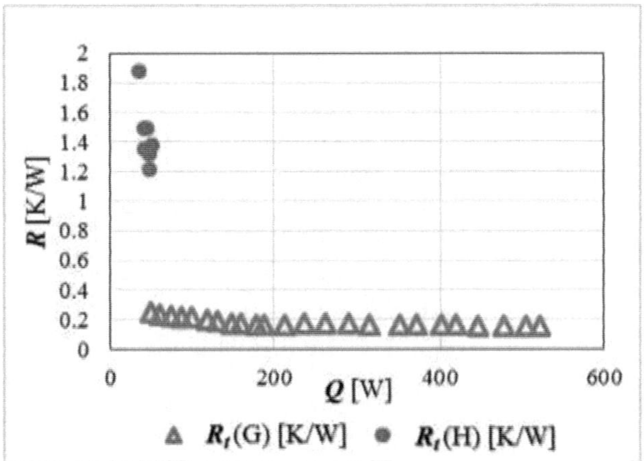

Figure 7. Rt of gravity assisted (G) and horizontal (H) LHP with heat load

4. Conclusions

In this study, a LHP with flat rectangular-shape evaporator is manufactured and compared the performance between horizontal condition and gravity assisted condition.

It is confirmed that circulation happens in both horizontal and gravity assisted condition. However, heat transfer capacity of horizontal LHP is lower than gravity assisted LHP because of higher pressure loss. Temperature at the surface of heater Ts1 is higher than 85 0C and total thermal resistance is greater than 1.2 K/W.

The gravity assisted LHP can operate stably in the range of heat load from 50 W to 520 W when temperature at the base of evaporator is lower than 100oC. Besides, this LHP can keep Ts1 below 85oC, which can be regarded as limitation temperature for the reliable and effective operation of electronics, when operating at heat load around 350 W. The operation characteristics of LHP can be classified into two modes that are variable thermal resistance and constant thermal resistance.

References

[1] Murshed S and Castro N D 2017 *Renew Sustain Energy Rev.* **78** 821
[2] Zhou G, Li J and Lv L 2016 *Appl Therm Eng.* **109** 514
[3] SUZUKI, Ayaka (Department of Advanced Mechenical Systems KU, SATO, Kaoru (Corporate R&D Division, Panasonics Ecology Systems Co, Ltd, Aichi J, KOITO, Yasushi (Department of Advanced Mechenical Systems KU, TOMIMURA, Toshio (Department of Advanced Mechenical Systems KU. Prototype Experiment on Cooling Performance of a JEST-

type Loop Heat Pipe. 4th World Conf Appl Sci Eng Technol. 2015;(October):24–7.
[4] Kondou C, Umemoto S, Koyama S and Mitooka Y 2017 *Appl Therm Eng* **118** 147–58
[5] Shioga T and Mizuno Y 2015 *Annu IEEE Semicond Therm Meas Manag Symp.* 50
[6] Ebrahimi K, Jones G F and Fleischer A S 2014 *Renew Sustain Energy Rev* **31** 622–38

Nomenclature

(a) Symbol	(b) Meaning	(c) Unit
OD/ID	Pipeouter, innerdiameter	(m)
	Copper thermal	
k	conductivity	(W/m·K)
	Heated area	
A	Heat flux	(m^2)
q	Heat load	(W/m^2)
Q	Total thermal resistance	(W)
R_t	Distance among	(K/W)
δ	thermocouples	(m)
	Surface tension	
σ	Pore radius	(N/m)
r	Heater temperature	(m)
T_1, T_2, T_3	Evaporator base	(°C)
T_4	temperature	(°C)
	Room temperature	
T_a	Temperature at condenser	(°C)
T_{co}	outlet	(°C)
	Temperature at	
T_{cci}	compensation chamber	(°C)
	inlet	
	Temperature at	
T_{eo}	evaporator outlet	(°C)
	Temperature at heater	
T_{s1}	surface	(°C)

IOP Conf. Series: Journal of Physics: Conf. Series **1086** (2018) 012014 doi:10.1088/1742-6596/1086/1/012014

Evolution of physical properties in hole doped Sr_2IrO_4: a J_{eff} = 1/2 magnetic insulator

I N Bhatti and A K Pramanik

School of Physical Sciences, Jawaharlal Nehru University, New Delhi-110067, India

E-mail: inbhatti07@gmail.com

Abstract. Recently 5*d* transition metal oxides (TMOs) particularly iridates have received extensive attention of researchers due to potential for exotic physics. Sr_2IrO_4 is extensively studied iridate over the past several years. Sr_2IrO_4 received great attention of researchers due to structural similarities with parent high T_C cupric superconductor La_2CuO_4. The layered perovskite Sr_2IrO_4 believed to stabilize in J_{eff} = ½ ground state and it is a magnetic insulator. The Sr_2IrO_4 shows weak ferromagnetic ordering below ~225 K, originating from the canted anti-ferromagnetic arrangement of the J_{eff} = 1/2 moments. Sr_2IrO_4 become interest compound for experimental research since it has been theoretically predicted to be superconductor upon electrons/hole doping. In this paper, we endeavour to tune SOI, U and introduce hole in electronic band by substituting Ru^{4+} ($4d^4$ S = 1) for Ir^{4+} ($5d^5$ J_{eff} = 1/2) in Sr_2IrO_4. We investigated the effect of Ru doping on structural and magnetic properties of Sr_2IrO_4. We observe evolution of both *a* and *c* lattice parameters in the meantime the octahedral distortion (θ_{Oct}) also decreases with doping. Further, our magnetization study shows suppression in transition (T_C), hysteresis with Ru doping. We found introduction of hole in Sr_2IrO_4 prominently effect the structural and magnetic properties.

1. Introduction

In 5*d* iridates active interactions such as: crystal field effect (CFE), on site coulomb interaction (U) and spin orbital coupling (SOC), compete among each other create new balance of interacting energies which promote novel electronic and magnetic phases in these materials. The most studied iridate layered perovskite Sr_2IrO_4 believed to stabilize in novel J_{eff} = ½ ground state it is a magnetic insulator driven by strong SOC. It is recently found that the electronic ground state is influenced by strong SOC of Ir ion in an octahedral environment which splits the t_{2g} band into lower $J_{eff} = 3/2$ quartet and upper $J_{eff} = 1/2$ doublet. The $J_{eff} = 1/2$ band is so narrow that modest U opens a Mott gap by splitting it into upper and lower Hubbard bands, thus driving it to unconventional Mott insulator.[1] The establishment of novel $J_{eff} = 1/2$ ground state has induced exotic properties in iridates. Sr_2IrO_4 crystallize in reduced tetragonal structure with space group $I4_1/acd$. The reduced symmetry is due to tiling of IrO_6 octahedral around c-axis. [1-7] Further, the structural organization of Sr_2IrO_4 is quite interesting and play crucial role in deciding physical properties of this material. The magnetic ground state in Sr_2IrO_4 is canted type antiferromagnetic (AFM) which gives rise to ferromagnetic component with Transition temperature (T_C) around 225 K. The spin canting nature is give rise to anti symmetric Dzyaloshinskii-Moriyam interaction driven by SOI and IrO_6 octahedra distortion. [3-8]. The ground state in this material is result of delicate balance of interacting energies viz, SOC, CFE and U, since tuning the strength of these energies can tune the physical properties. Motivated by this fact we adopt chemical doping of Ru^{4+} ($4d^4$) for Ir^{4+} ($5d^5$) this will tune both SOC and U, moreover introduce a hole into the band. Tuning of this crucial perimeter will perturb the ground state and hence evolve physical properties of this compound. [5-10]

In this paper we investigate the evolution of structural and magnetic properties of Sr_2IrO_4 with Ru doping. The structural characterization is done with X-ray diffraction. We observe evolution of lattice parameter with Ru doping. The octahedral distortion which is crucial in deciding physical properties reduces with doping and <Ir-O-Ir> bond angle increases. Both temperature and field dependent magnetization is studied to understand the magnetic properties. We observe sample undergo paramagnetic (PM) to ferromagnetic (FM) phase transition and suppression of T_C with reduce magnetic moment. However, the effective paramagnetic moment increases and Curie temperature decreases. The isothermal magnetization M(H) shows decrease in coercively and remnant magnetization.

2. Experimental Detail

The polycrystalline samples of $Sr_2Ir_{1-x}Ru_xO_4$ (x = 0.0 and 0.1) were prepared using standard solid state method using ingredients $SrCO_3$ and IrO_2 and RuO2 from Sigma Aldrich. The high purity ingredients were thoroughly mixed and grinded well. The fine ground powder then calcinated at 900 ^{0}C for 24 hour with heating and cooling rate set to 3^0C/min. The calcinated powder is further grounded for half an hour and pressed into pallet and sintered at 1000 ^{0}C and 1100 ^{0}C for 24 hours each with intermediate grindings to achieve sample homogeneity. The sample preparation procedure described in detail elsewhere.[8-12] The phase purity of the samples has been checked using powder x-ray diffraction (XRD) with a Rigaku MiniFlex diffractomer with CuKα radiation. The XRD data is used for structural analysis with the help of Full Proof software. Both temperature and field dependent DC magnetization (M) data have been collected using a vibrating sample magnetometer (PPMS, Quantum Design).

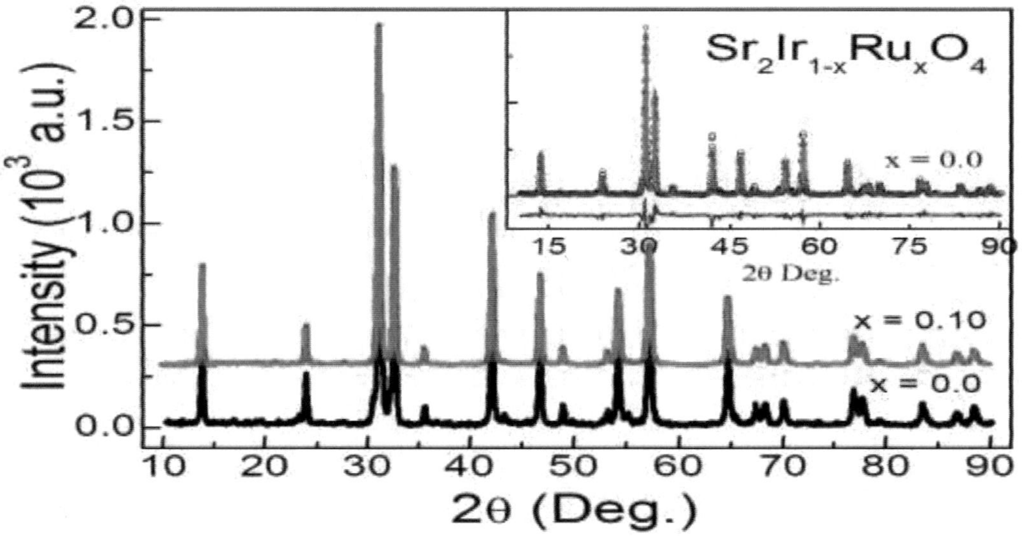

Figure 1. XRD pattern collected for $Sr_2Ir_{1-x}Ru_xO_4$ has been shown along with the Reitveld refinement.

3. Result and Discussions

3.1. Structural study

XRD patterns of $Sr_2Ir_{1-x}Ru_xO_4$ sample series are shown in Fig. 1, the comparison of XRD patterns shows no extra peak evolves with doping. Inset Fig. 1 shows Rietveld refinement analysis of parent compound. The structural analysis shows that all samples are in single phase and chemically pure. All samples crystallize in tetragonal phase with $I4_1/acd$ space group. The refined lattice parameters are $a = b$ = 5.4980(2) Å and c = 25.779(1) Å for x = 0.0 sample, both a and c lattice parameters decreases with Ru doping. We also find a rotation of IrO_6 octahedra decreases from ~ 11.3° around c axis for parent compound to 9° for x = 0.1 sample.

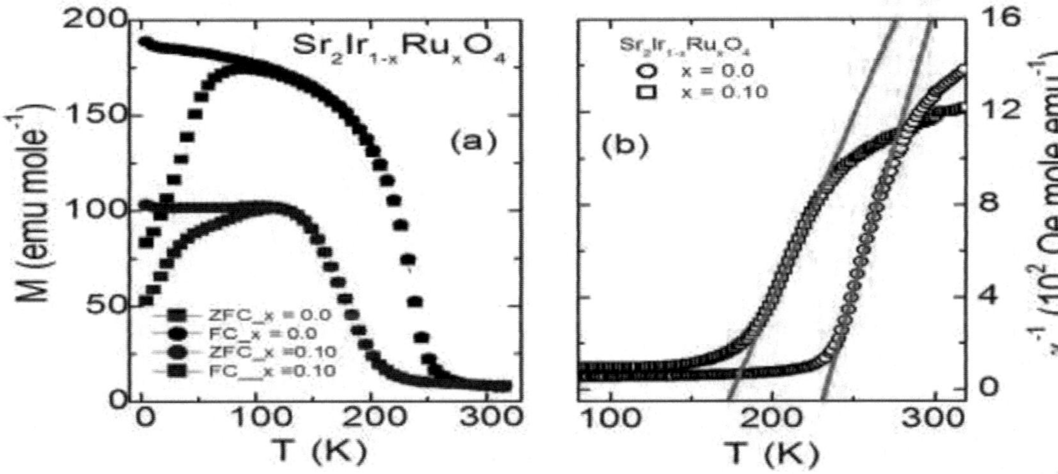

Figure 2. (a) ZFC and FC magnetization data measured in 10 kOe are plotted as a function of temperature for $Sr_2Ir_{1-x}Ru_xO_4$. (b) Inverse susceptibility (χ^{-1}) shows Curie-Weiss behaviour.

3.2 Magnetization Study

Fig. 2(a) shows the temperature dependant magnetization M(T) data collected in ZFC and FCW protocol in the temperature range of 5 K – 300 K in an applied field of 10 kOe. The M(T) data show a phase transition from paramagnetic (PM) to weak ferromagnetic (FM) state with T_C~ 225 K for parent compound. However, with Ti doping T_C decreases. Fig. 2(b) shows the invers susceptibility as a function of temperature. In PM regime the magnetic susceptibility is fitted with Curie-Weiss law expressed as:

$$\chi = \frac{C}{T - \theta_P}$$

Where χ is magnetic susceptibility, C is the Curie constant and θ_P is the Curie temperature. We calculated effective paramagnetic moment for all samples using fitting parameters. We found the μ_{eff} = 0.56 μ_B/f.u for parent compound and μ_{eff} increases with Ru doping

To further understand the magnetic properties we measured isothermal magnetization M(H) at T = 5 K shown in Fig. 3(a). The hysteresis in M(H) is suggestive of FM ordering in this material. At 5 K, the magnetic moment μ_S = 0.05 μB/f.u at H = 70 kOe is observed, which decreases with Ti doping. It is quite evident from the Fig. 3(a) that corecivity and reminance decreases with doping. Fig. 3(b) shows the Arrott plot (M^2 vs H/M) ,the extrapolation of Arrott plot at high field intercept the positive M^2 axis which realize the presence of spontaneous moment in these samples.

The suppression of transition temperature TC and moment can be explain by considering the effect of Ru substitution. Since Ru^{4+} is S = 1 and Ir4+ is Jeff =1/2 thus Ir-O-Ru-O-Ir interaction is weaker then Ir-O-Ir-O-Ir, due to with T_C decreases and system become less ferromagnetic.

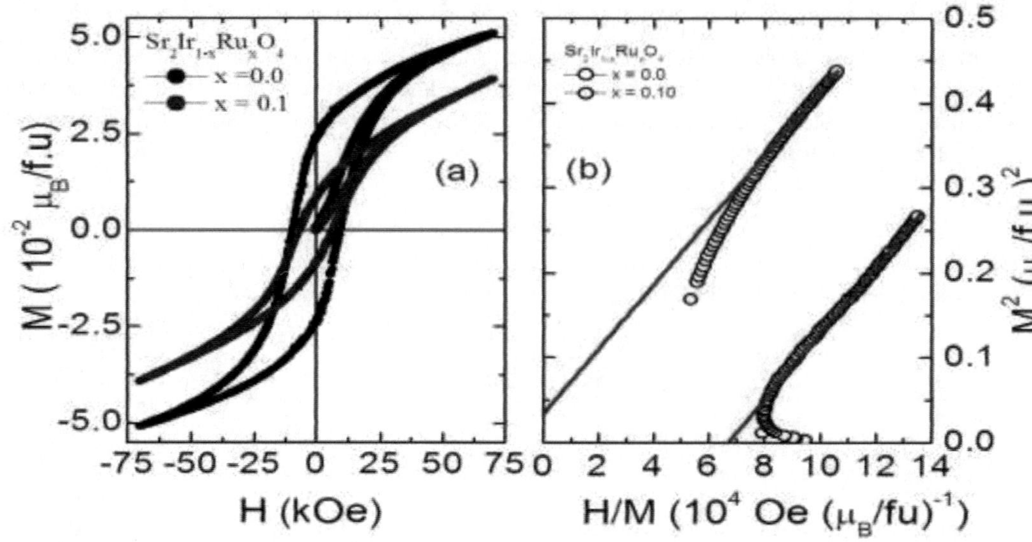

Figure 3. (a) Isothermal magnetization is shown for $Sr_2Ir_{1-x}Ru_xO_4$ at 5 K. (b) Fitting of M(T) following Eq. 1 is shown.

4. Conclusions

In conclusion, we successfully prepared the single phase high purity polycrystalline samples of $Sr_2Ir_{1-x}RuO_4$. Structural analysis reveals that samples crystalize in tetragonal structure with reduced symmetry. With Ru doping lattice parameters evolve and octahedral distortion reduces. The magnetization study shows samples undergo magnetic phase transition from PM to FM. Ru doping result in enhancing effective magnetic moment and decreases curie temperature. The M(H) data shows hysteresis which decreases with Ru doping. To further understand the magnetism in Ru doped Sr_2IrO_4 samples with even higher Ru doping need to be prepared and studied.

5. Acknowledgments

We acknowledge UGC-DAE CSR, Indore, India for measurement facilities. One of the authors (INB) acknowledges the CSIR, India for fellowship.

References

[1] Kim B J, Jin H, Moon S J, Kim J -Y, Park B -G, Leem C S, Yu J, Noh T W, Kim C, Oh S -J, Park J H, Durairaj V, Cao G, and Rotenberg E 2008 *Phys. Rev. Lett.* **101** 076402

[2] Moon S J, Jin H, Kim K W, Choi W S, Lee Y S,Yu J, Cao G, Sumi A, Funakubo H, Bernhard C, and Noh T W 2008 *Phys. Rev. Lett.* **101** 226402

[3] Crawford M K, Subramanian M A, Harlow R L, Fernandez-Baca J A, Wang Z R, and Johnston D C 1994 *Phys. Rev. B* **49** 9198

[4] Ge G,Qi T F, Korneta O B, De Long D E, Schlottmann P, Crummett W P, and Cao G M 2011 *Phys. Rev. B* **84** 100402

[5] Lee J S, Krockenberger Y, Takahashi K S, Kawasaki M, and Tokura Y 2012 *Phys. Rev. B* **85** 035101

[6] Shimura T, Inaguma Y, Nakamura T, Itoh M, and Morii Y 1995 *Phys. Rev. B* **52** 9143

[7] Calder S, Cao G -X, Lumsden M D, Kim J W, Gai Z, Sales B C, Mandrus D, and Christianson A D 2012 *Phys. Rev. B* **86** 220403

[8] Bhatti1 I N, Rawat R, Banerjee A and Pramanik A K 2014 *J. Phys.: Condens. Matter* **27 (1)** 016005

IOP Conf. Series: Journal of Physics: Conf. Series **1086** (2018) 012014 doi:10.1088/1742-6596/1086/1/012014

[9] Bhatti I N and Pramanik A K 2017 *J. Magnetism and Magnetic Material* **422** 141-148
[10] Bhatti I N and Pramanik A K 2016 *AIP Conference Proceedings* **1731** 110046
[11] Bhatti I N and Pramanik A K 2018 *AIP Conference Proceedings* **310** 012055
[12] Bhatti I N, Dhaka R S, and Pramanik A K 2017 *Phys. Rev. B* **96** 144433

IOP Conf. Series: Journal of Physics: Conf. Series **1086** (2018) 012015 doi:10.1088/1742-6596/1086/1/012015

Microstructure and electric properties of spray pyrolizedCdO thin films grown on SILAR deposited CdO seed

R Biswas[1], M M Rahman[1], M K R Khan[1] and M Kamruzzaman[2]

[1]Department of Physics, University of Rajshahi, Rajshahi-6205
[2]Department of Physics, Begum Rokeya University, Rangpur 5400

E-mail: etithuru36@gmail.com

Abstract. In this report, a substantial effect of seed layer on structural and electrical properties of CdO thin film have been presented. Thin films of CdO have been synthesized by spray pyrolysis method at 400 °C on seed layer of CdO grown by SILAR method on glass substrate at 85 °C. XRD study revealed a mono-phasic cubic structure of CdO thin films oriented along (2 0 0) plane with highest intensity when grown on seed layer. Electrical conductivity, Hall mobility, carrier concentration are found substantially high using seed layer which suggests that seed layer plays a crucial role to enhance better physical and electric properties of CdO films.

1. Introduction

There is an ever growing interest in the synthesis of low-cost, transparent conducting oxides (TCOs) with n-type conductivity [1]. CdO is an n-type semiconductor having band gap between 2.2 and 2.7eV [2] and shows low resistivity due to defects of oxygen vacancies and cadmium interstitials [3]. In search of such suitable TCOs, it is found that chemical and physical conditions can be conveniently incorporated in the CdO thin films to obtain higher transmission and conductivity [1]. Cadmium oxide (CdO) thin films are regarded as a material with as large energy band gap, high optical transparency in the visible spectral region, remarkable luminescence characteristics etc. Due to these properties, CdO is a promising material for electronic or optoelectronic application, such as solar cells application, photodiodes and gas sensors [4]. For the increase of solar cell efficiency it is necessary to reduced energy loss in electrodes. Therefore, resistivity of electrodes should be as low as possible. In a good crystalline film, it is expected, defect will minimum and thereby carrier scattering will also be minimum. As a result, resistivity will be low in a good crystalline film. For obtaining low resistive and good crystalline film, crystalline seed layer was used on glass substrate for preparation of CdO film.

Up to now, various techniques such as sputtering [5, 6], vacuum thermal evaporation [7] pulsed laser deposition (PLD) [8], chemical vapor deposition (CVD) [9], successive ionic layer adsorption and reaction (SILAR) [10, 11], sol–gel [12,13], spray-pyrolysis (SP) [14], chemical bath (CB) [15] and so on are employed to grow CdO thin films. Among these, spray pyrolysis technique is simple, safe and low-cost, but effectively capable of producing film of high aspect ratio with desired crystalline quality. Moreover, in SP method excellent control of chemical uniformity, stoichiometry in multi-component system and varying band gap materials processing are possible in air ambient [16-19].

In this work, we present a new approach, a combination of spray and spray on seed layer (seed prepared by SILAR technique) method called 'spray on seed layer growth' to fabricate CdO thin films.

First, a very thin layer (seed layer) of CdO on glass substrate was deposited by SILAR technique and then spray was made on the seed layer from the solution of $(CH_3COO)_2Cd.\ 2H_2O$at 400 °C. As-grown films were studied to observe the effect of seed layer on the structure and electric properties of CdO thin films.

2. Experimental

A very thin seed layer of CdO has been deposited on glass substrate by SILAR method, over which CdO thin films have been prepared by spray pyrolysis method at 400 °C. For this purpose cadmium acetate di-hydrate ($(CH_3COO)_2Cd.\ 2H_2O$) (Merck 99.9%) was used as precursor for Cd^{2+} ion source and H_2O_2 for anion source.

For obtaining a good seed layer of CdO on glass substrate by SILAR method it is necessary to clean substrate properly, otherwise adhesion of seed on glass would be poor. As a result seed layer will be washed out during spray. The following protocol was adopted for substrate cleaning.

i) Glass substrates were dipped into detergent about 24 hours, cleaned thoroughly in a stream of cold water and then washed by the de-ionized distilled water.
ii) The substrates were then dipped into nitric acid for 1 hour and then washed thoroughly in distilled water and then dried.
iii) These substrates were then dipped into piranha solution (solution of H_2SO_4 and 30% H_2O_2 in a proportion of 1:3) for one hour and then held in a stream of cold water for some times followed by rinsing with de-ionized distilled water several times. Finally, they were dried by blowing hot air with the help of drier and preserved in a slide holder for use.

a) Preparation of seed layer: CdO seed layer was prepared by a four-step SILAR technique using a solution comprising 0.1 M cadmium acetate ($(CH_3COO)_2Cd.\ 2H_2O$), 25% ammonia solution with a pH value 11.5 at room temperature. Hydrogen peroxide (H_2O_2) solution was used as the source of anion. For CdO seed preparation, following four steps were adopted:

STEP-1: At first 0.1 M cadmium acetate ($(CH_3COO)_2Cd.\ 2H_2O$) was taken in a beaker and ammonia solution (25%) was instantly poured into it. The pH value of this solution was fixed at 11.5. The well-cleaned substrates were inserted into this solution for 30 seconds so that the Cd^{2+} ion adhered onto the glass substrate and form a very thin layer of $Cd(OH)_2$, shown in figure 1(a).

STEP-2: This $Cd(OH)_2$layer was washed with distilled water in order to remove loosely bound cationic particles. Then the substrates were dipped into warm distilled water ((85 ± 5) °C) about 10 to 15 seconds, shown in figure 1(b).

STEP-3: Immediately after step-2, substrates were dipped into a cationic source hydrogen peroxide (H_2O_2) solution (0.05M) for 30 seconds for anion uptake on the substrate. The complex layer deposited on the substrate during the dipping in hydrogen peroxide (H_2O_2) bath will be decomposed to CdO, shown in figure 1(c)].

STEP-4: Finally, the substrate was dipped into hot water (85 ± 5) °C for 10 to 15 seconds followed by withdrawal of substrate and dried. These four steps together with complete a full cycle, as shown in figure 1(d)].

The possible chemical reactions that take place on the glass substrate at different stages during formation of CdO seed are given bellow:

$$(CH_3COO)_2Cd.\ 2H_2O + NH_4OH \longrightarrow Cd(OH)_2 + CH_3COONH_4 + H_2O$$

$$Cd(OH)_2 + NH_4OH \longrightarrow (NH_4)CdO_2^- + H_2O + H^+$$

$$(NH_4)CdO_2^- + H^+ \longrightarrow CdO + NH_4OH$$

b) Spray on seed: In spray pyrolysis, condensation of atoms or molecule onto a heated substrate forms a thin film of CdO. Spray deposited film was formed using 0.01M cadmium acetate [$(CH_3COO)_2Cd$. $2H_2O$] by conventional spray technique [20]. For obtaining a good quality film, optimized growth conditions were nozzle to substrate distance 21 cm, spray rate 0.3 ml/min, spray time 20 min. and substrate temperature 400 °C.

The chemical reaction that takes place on the heated substrate in spray method is given as:

$$(CH_3COO)_2Cd.\,2H_2O \xrightarrow{\text{Decomposes at } 400\,^\circ C} CdO + CO_2\,(g) + CH_4\,(g) + \text{Steam}\,(g)$$

(a) $(CH_3COO)_2Cd.2H_2O$+ NH_4OH **(b) H_2O at 85 °C** (c) H_2O_2 **(d) H_2O at 85**

Figure 1. Schematic representation of SILAR method (a) cationic precursor and (c) anionic precursor and (b, d) de-ionized water.

An X- Ray diffractometer (D2 PHASER BRUKER) was used to study the structural properties of CdO films deposited on glass substrate by SILAR, spray and spray on seed layer growth techniques. The X-ray diffraction of all samples has been recorded within Bragg angle (2θ) of 20 ° to 70 ° using monochromatic CuKα radiation of wavelength λ = 0.1542 nm, from a source power 30 kV-10 mA. Temperature dependent electrical conductivity and RT Hall voltage measurements were performed by Vander Pauw's method using Keithley electrometer (England). Substrate temperature was controlled directly by adjusting voltages in a digital multi-meter (KT105) Temperature was monitored by a copper-constantan thermocouple. For Hall field, an electromagnetic type magnetometer (Newport instruments Ltd. England) was used at constant magnetic field of 0.9815 Tesla. Thicknesses of the films were (180 ± 10) nm measured by Newton's ring method.

3. Result and discussion

3.1 Structural Properties

The physical properties of thin films are highly related to their structure. A typical XRD pattern of CdO thin films prepared by SILAR, spray and spray on seed layer growth are shown in figure 1. In this figure, the presence of several peaks in the XRD patterns reveal that all the films are polycrystalline and characteristic reflection peaks are assigned to the (111), (200), (220), (311) planes by comparing with the standard, JCPDS card No. 75-0594 **[21]**. From XRD patterns, it is seen that films are polycrystalline in nature and have cubic CdO single phase. The highest intensity peak of CdO was produced by SILAR technique in (111) plane but for spray and spray on seed layer growth techniques, it was in (200) plane. It has also been observed that CdO film on seed layer showed highest intensity of reflections in (200) plane. The highest peak intensity and sharpness of the peak indicate better crystallinity of the films. The film grown on glass substrate by SILAR technique has

[111] preferred orientation, which is energetically favorable for CdO observed by other methods too, but for spray and spray on seed growth, preferred orientation changed to [200]. This change of orientation indicates that growth of CdO film at high temperature is energetically favorable in [200]. It seems the microstructure of CdO lattice changes due to growth on seed layer which inevitably perturbed the lattice thermodynamic energy of the growth planes, thereby, at high temperature growth of CdO on seed layer [200] plane is energetically more favorable.
The crystallite size (ζ) was estimated by using the Debye-Scherrer formula **[22]**,

$$\zeta = \frac{K\lambda}{\beta \cos\theta} \quad (1)$$

where, K is a constant close to unity (0.94). The parameter λ is the wavelength of the X-ray used, θ is the diffraction angle and β is the full width at half maxima (FWHM) measured in radians. The dislocation density (δ), defined as the dislocation lines per unit volume of the crystal, has been calculated by using the formula **[23],**

$$\delta = \frac{1}{\zeta^2} \quad (2)$$

The lattice strain (η) is calculated using the relation **[23],** $\eta = \frac{\beta}{4\tan\theta}$ and the estimated values are presented in Table 1.

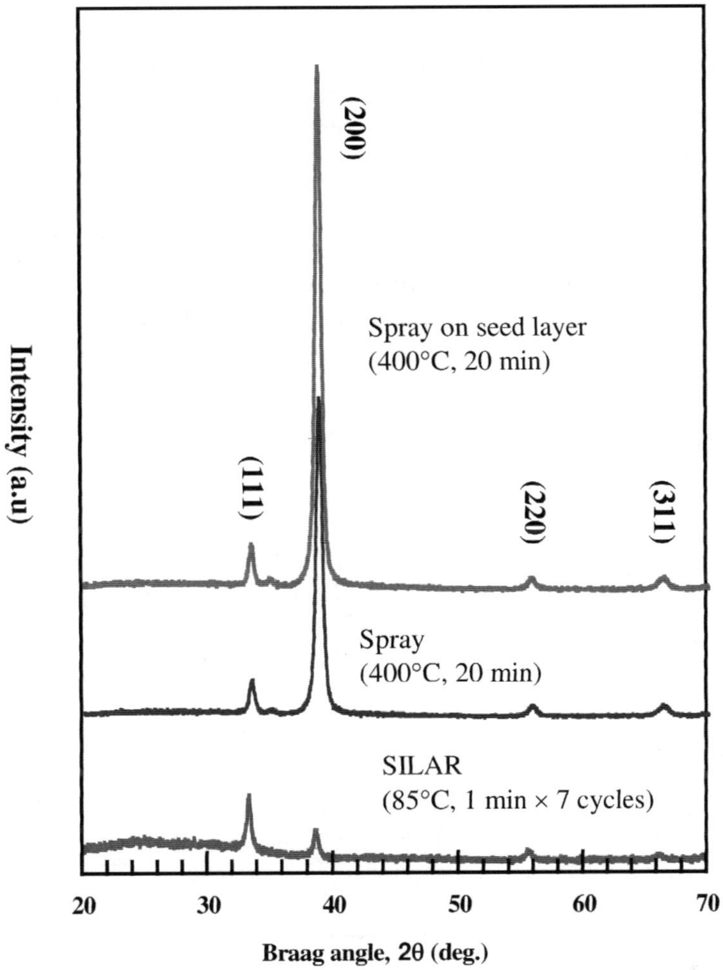

Figure 2. XRD pattern of CdO thin films

IOP Conf. Series: Journal of Physics: Conf. Series **1086** (2018) 012015 doi:10.1088/1742-6596/1086/1/012015

Table 1. Estimated lattice parameter, crystallite size, strain and dislocation density of CdOthin film for (200) plane.

Deposition method	Lattice parameter, a (nm)	Crystallite size, ξ (nm)	Lattice strain, $\eta \times 10^{-3}$	Dislocation density, $\delta \times 10^{-3}$ (nm)$^{-2}$
SILAR	0.4660	34	1.95	0.32
spray	0.4573	24	4.79	1.99
spray on seed	0.4634	23	4.38	1.62

3.2 Electrical Properties

The Vander Pauw's electrical conductivity with temperature for CdO film is shown in figure 3. There is a small upturn of conductivity from 300 K to 330 K noticeable for sprayed film, which is ascertained as poor metallic nature [20] of CdO film due to degeneracy, after which normal semiconducting behaviour was observed. On the other hand, spray on seed thin film shows semiconducting behaviour all through the measured temperature range. However, both films show very high conductivity but spray on seed film shows about four time higher conductivity compared to sprayed film which would make spray on seed film more efficient as solar cell electrode.

Hall voltage measurement at room temperature results n-type carrier concentration of the order of ~10^{20}/cm^3 but the pre-exponential term is more than four times higher for spray on seed layer growth compared to spray growth alone. However, mobility almost same but substantially higher compared to other growth methods, and also the carrier concentration and RT conductivity are much higher than other studies as presented in Table 2 for comparison. All these transport parameters of this study show superior to other growth technique **[2, 9, 24]**, which are costly and need vacuum environment.
The activation energy of CdO thin film was estimated from the relation,

$$\sigma = \sigma_o exp\,(-\Delta E/2k_BT) \quad (3)$$

where ΔE is the activation energy, $k_B$is the Boltzmann constant and $\sigma_o$is the pre-exponential term. Inset of figure 3 shows the plot of lnσ vs. 1000/T for CdO thin films fabricated by spray and spray on seed layer growth methods. The slop of the curves gives the activation energy. Activation energy for sprayed film is 0.043 eV and 0.003 eV for spray on seed film. These activation energy values suggest electrical transport is thermally governed in CdO thin film.

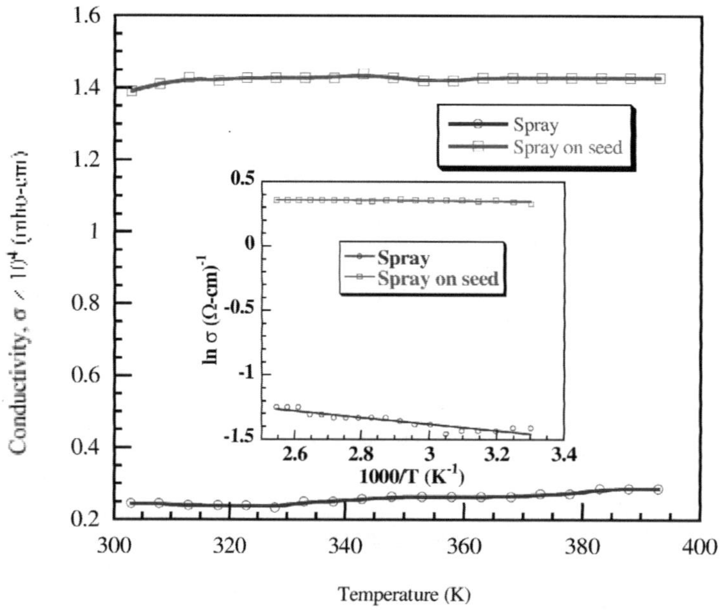

Figure 3. Variation of electrical conductivity with temperature and lnσ versus 1000/T (inset) graph

Table 2. Hall mobility, carrier concentration, RT conductivity and activation energy of CdO thin film.

Deposition method	Hall mobility μ_H (cm^2/V-sec)	Carrierconcentration $\times 10^{20}$ $(cm)^{-3}$	RTConductivity $\sigma \times 10^4$ (mho-cm)	Activation energy, ΔE (eV)	Ref.
Spray	33	4.5	0.24	0.043	this work
Spray on seed	34	20.4	1.11	0.003	this work
MOCVD	32	0.24	0.026	-	[9]
SGDC	8.0	0.39	0.005	-	[24]
TE	7.03	4.43	0.005	-	[2]

4. Conclusions

Spray and spray on seed layer growth CdO thin films were prepared to observe the effect of seed layer on the physical properties of CdO films and the following conclusions are made:

i) CdO thin films prepared by spray and spray on seed layer are mono-phasic having cubic rock salt structure; preferred orientation [111] for seed layer grown by SILAR technique changed to [200] when CdO sprayed on seed at high temperature. Spry on seed layer growth technique provide structurally better film over spray method because of CdO seed layer on glass substrate acts as a building block for crystallization.

ii) Very high RT conductivity, $\sigma = 1.1\times10^4$ (mho-cm), high mobility, $\mu_H = 34$ (cm^2/V-sec) and high carrier concentration, n $=20.4\times10^{20}$ $(cm)^{-3}$ were obtained for CdO film prepared by spray on seed growth technique. High values of these transport properties suggest CdO film prepared by spray on seed layer would be more efficient for solar cell electrode and opto-electric device applications.

iii) Activation energy values suggest electrical transport is thermally activated in CdO film.

Acknowledgements

One of the authors R. Biswas is thankful to the Ministry of Science and technology (MOST) of the Peoples Republic of Bangladesh for providing financial support.

References

[1] Yusuf G T, Babatola B K and Ishola A D 2016 *Science and Engineering Applications* **1(7)** 92-95
[2] Dakhel A A 2008 *Semicond. Sci. Tech.***23** 055017
[3] Zhao Z, Morel D L and Ferekides C S 2002 *Thin Solid Films* **413**, 203
[4] Naser G Y, Raja W N, Faris A S, Rahemd Z J, Salihe M A and Ahmed A H 2013 *Energy Procedia* **36** 42- 49
[5] Zhou Q, Ji Z, Hu B, Chen C, Zhao L and Wang C 2007*Mater. Lett.* **61** 531
[6] Saha B, Thapa R and Chattopadhyay K K 2008 *Sol. Energy Mater., Sol. Cells* **92**, 1077
[7] Suhail M H, Ibrahim I M and Mohan Rao G 2012 *J. Electron. Dev.* **13** 965
[8] Yan M, Lane M, Kannewurf C R and Chang R P H 2009 *Appl. Phys. Lett.* **78** 2342
[9] Lamb D and Irvine S J C 2009 *Thin Solid Films* **518** 1222
[10] Salunkhe R R and Lokhande C D 2008 *Sens. Actuators* **B129**, 345
[11] Ergin B, Ketenci E and Atay F 2009 *Int. J. Hydrogen Energy* **34** 5249
[12] Ayala A M, Delgado G T, Perez R C and Angel O Z 2010 *J. Alloys Compd.* **506** 554

[13] Ilican S, Caglar M, Caglar Y and Yakuphanoglu F 2009 *Optoelectron. Adv. Mater. RapidCommun.* **3** 135
[14] Jeyaprakash B G, Kesavan K, Kumar R A, Mohan S and Amalarani A 2010 *J. Am. Sci.* **6** 75
[15] Al-Ogili H K J 2011 *Eng. Technol. J.* **29** 1536
[16] Rao T P, Kumar M C S, Angayarkanni S A and Ashok M 2009 *Journal of Alloys and Compounds* **485** 413
[17] Babu B J, Maldonado A, Velumani S and R Asomoza 2010 *Materials Science andEngineering* ***B*174** 31
[18] Manouni A E*et al.* 2006 *Superlattices and Microstructures* **39** 185
[19] Rahman M M*et al.* 2012 *Journal of Materials Science & Technology* **28** 329
[20] Khan M K R, Rahman M A, Shahjahan M, Rahman M M, Hakim M A, Saha D K and Khan J U 2010 *Curr.Appl. Phys.***10** 790-796
[21] Usharani K and Balu A R 2015 *Journal of ActaMetall.Sin., Engl.* **28(1)** 64-71
[22] Cullity B D 2001 *Elements of X-ray diffraction,* Prentices Hall, New Jersey **388** 619
[23] Yilmaz M, Aydin S, Turgut G, Dilber R and Ertugrul M 2012 *Progresses in Nanotechnology and Nanomaterials* **1(1)** 5-8
[24] Murali K R, Kalaivanan A, Perumal S and Pillai N N 2010 *J. Alloys Compd.* **503** 350-353

Institute of Physics
Dirac House, Temple Back
Bristol BS1 6BE UK

ISSN: 1742-6588
ISBN 978-1-5108-7513-5

CURRAN ASSOCIATES INC.
proceedings.com